Praise for Think Like an E

"I am struck by the similarities between the approach (de~~~~~~~~ ~~ ~~~ ~~~~) and the outcomes we have seen in Yale's interdisciplinary Center for Engineering Innovation and Design, where artists and business students work alongside engineers."

Prof. Peter Salovey, President, Yale University

"The word 'engineer' has the same roots as the word 'ingenious' (Latin 'ingenium'). Engineers are therefore charged with finding ingenious solutions to mankind's problems. This book describes how to think like an engineer – using creative formal and informal approaches - and how to arrive at optimal solutions using the Conceive, Design, Implement, Operate (CDIO) process. It will be a useful text for undergraduate engineers wanting to design even more ingenious solutions!"

Prof. John Fothergill, Pro Vice-Chancellor, City University London

"This book is a real tour de force. Prof Al-Atabi approaches his topic 'Think like an engineer' from the very foundations of how our human brain operates, and interacts with our senses, how we build models of our world and use these models to guide us to realise (and when necessary limit) our dreams.
'Dream Big. Be Different. Have Fun.' is a very appropriate chosen by-line for the book. Indeed the book itself is an example of how one can live this experience. Reading the book, one feels certain that Prof Al-Atabi was doing just that dreaming big, being different and having fun.
For everyone who wants to be an engineer, this book will be a great guide to unleash their creativity – after all, that is the true meaning of the Latin "ingenerare" (to create) from which the English word engineer is derived. Have fun reading!"

Prof. Iven Mareels, Dean, Melbourne School of Engineering

"This book articulates that the thought process of an engineer is not just about specific technical accomplishments, but about acquiring the skills to handle the evolution of emotions, goals, and the environment in a dynamic world. It broadens the scope of engineering practice as a principle, not merely a profession. A must read for citizens in this technological world!"

Prof. Ben Koo, Tsinghua University, Beijing, China

"As an experienced academic who leads first year engineers and complex research teams across academia and industry, this book resonates closely with many of my teaching and learning experience. The engineering CDIO (Conceive-Design-Implement-Operate) process is cleverly interlaced with relevant and innovative concepts in brain function biochemistry, positive thinking, emotional intelligence, brainstorming, project management, and entrepreneurship. This creates a much more holistic view of engineering thinking, and will influence future course designs in my University"

Prof. Pete Halley, Head of School of Chem. Eng., University of Queensland

"Project-based learning is essential and valuable to engineers and engineering students who are expected to apply scientific knowledge, mathematics, and ingenuity to resolve technical, societal, and commercial problems in order to benefit the human beings in this century. Its core elements are all addressed in the book 'Think Like an Engineer'."

Dr. Yuei-An Liou, Distinguished Professor,
National Central University, Taiwan

"'Think Like an Engineer' moves well beyond the stereotypical view of Engineers as mathematical thinkers, Al-Atabi highlights the importance of communication, emotional intelligence and empathy in solving problems for society. Illustrating his points with appropriate examples, the reader is taken through a multi-dimensional perspective on the mindset of a successful Engineer. What stands out is how accessible the concepts are - this book is easy to read and the key points are easy to grasp, but nonetheless the technical depth is preserved. An outstanding introduction to the Engineering profession, and one that is useful not just for potential engineers but to society at large."

Prof. Euan Lindsay, Foundation Professor of Engineering,
Charles Sturt University

"Engineers are entrusted with helping all of us face the Grand Challenges of the 21st Century. To achieve this, they need to think systematically and innovatively while searching for ground-breaking solutions. 'Think Like an Engineer' presents the CDIO framework of practice in an easy to read, down to earth manner that makes it accessible not only to engineers, but to the wide public as well. The book also provides a holistic personal and professional development programme as it incorporates the latest in brain science, emotional intelligence and entrepreneurship. It is a must read for engineers, business leaders and educators in general."

Prof. Seeram Ramakrishna, Founder of Global Engineering Deans Council, and
one of the World's Most Influential Scientific Minds

"Mushtak has tackled a subject close to my heart; he has explored what makes an engineer tick; the way in which engineers tackle real world and often global challenges; the deep down passion that keeps engineers motivated; and looks into the heart-ware needed for an eternally evolving discipline that is developing solutions for a sustainable future."

Joe Eades, Founder, Ispahan Group

"An invaluable book for both engineering students and those teaching them. It provides both an overarching conceptual rational for learning engineering –and related STEM disciplines –and practical frameworks to shape student understanding of their discipline."

Alan Jenkins, Professor Emeritus, Oxford Brookes University

"This lively engineering education book by Mushtak Al-Atabi is filled with easy to relate examples and a systematic treatise to support and enhance the effectiveness of the process."

Dr. Shuo-Yan Chou, Distinguished Professor,
National Taiwan University of Science and Technology

"The book is one of the best resource available for students interested in the topic. It is written in a straightforward and supportive style to guide readers in the CDIO concept of engineering projects. It provides clear, simple and practical illustrations as well as a variety of approaches, which enable readers to put into practice for their own use. This invaluable resource will go a long way in supporting students learning and preparing them for their engineering careers. Definitely worth reading."

Prof. Mohd Hamdi Abd. Shukor, Deputy Vice-Chancellor,
University of Malaya

'Think Like an Engineer' puts into perspective the underlying and at times the subconscious complexity of how engineers think. Many years of expertise and passion have gone into developing this text and it makes stimulating reading and is not constrained to the engineering profession. Whatever profession we have taken there is always a natural curiosity for finding out how things work and then improving them, this books amplifies the important ingredients that will allow that curiosity to be realised. Throughout the book the importance of the creative mind plays a key role in being able to generate ideas, artefacts, or processes that are new and of value and which make the world a better place. Failure is also recognised as a key part of being creative, if we are not making mistakes, we are not learning and we are certainly not doing things differently.

Professor Al-Atabi shares many years of expertise in a book that has evolved over time with his own experiences. It will be a key companion for engineering students who are engaged in project based activity and for professionals in any disciple who want to take a more structured and systematic approach to address the challenges that they face."

Professor Gary Hawley, Dean of Engineering, University of Bath

"Dr. Al-Atabi's book unlocks the conundrum of what to learn today to be relevant in the face of future challenges and opportunities that do not exist yet. He infuses his stories with an infectious passion to make an enjoyable read. Highly recommended to anyone who aspires to live a meaningful and productive life."

Arnold Teo, Shell (Malaysia)

"There are many books that talk about success and how to achieve it. What Prof. Mushtak has brought forward in this book is a unique view of how engineers think and what makes engineers successful. In summary, it was never more than following your passion relentlessly. I've always been a great fan of Prof. Mushtak and I'm absolutely pleased that he has taken this mammoth effort of documenting his lifelong journey of discovering what makes engineers tick."

Harvinder Singh, Chairman and Group Managing Director, PSI inCONTROL Sdn. Bhd.

"Engineers are known as problem-solvers. They apply science and maths to solve problems creatively. Mushtak Al-Atabi has captured the essence of engineering and made engineers more human by linking the brain and emotional intelligence. This well-written book provides coverage of a number of important issues and techniques not commonly covered in most introductory engineering textbooks. It would benefit both engineers and non-engineers to read it as it may help them solve problems or develop opportunities in the real world."

Mastura Mansor, Vice President-Global Supply Chain, Scomi

"The challenges of the 21st century will require a whole new kind of engineer: engineers who can both design complex technical systems and appreciate their interaction with society; engineers who possess both analytical intelligence and emotional intelligence; engineers who can work with others, take risks, and learn from failure. Al-Atabi's text provides engineering students with a thoughtful roadmap to grow and develop as this kind of engineer; students and faculty alike will benefit from his insights."

Prof. Mark Somerville, Olin College. Co-author of 'A Whole New Engineer'

This book provides a thoughtful connection between a hallmark engineering approach and the psychology of human decisionmaking. Using practical tips, summaries of important studies, and personal anecdotes, 'Thinking Like an Engineer' links relevant contemporary learning research to CDIO. This is a great resource for those interested in blending engineering thinking, global competence, emotional intelligence, and empathetic design.

Jennifer DeBoer, Asst. Prof. of Engineering Education, Purdue University

About the Author

Professor Mushtak Al-Atabi is currently the Provost and CEO of Heriot-Watt University Malaysia. A passionate educator, innovator and an agent of change, Mushtak always challenges the status quo to unlock value. He pioneered the use of the CDIO educational framework in Malaysia and offered one of the first Massive Open Online Courses (MOOCs) in Asia (the very first in Malaysia) in 2013 when he was the Dean of Engineering at Taylor's University. His online classes, *Entrepreneurship*, *Success with Emotional Intelligence* and *Global Entrepreneurship* attracted thousands of students from 150 different countries. He speaks at international conferences and consults for national and multinational corporations, including universities, banks and manufacturing and energy companies, in the areas of leadership, innovation, human development, performance and technology. Mushtak is the author of *Think Like an Engineer* and *Driving Performance* and the founding Editor-in-Chief of the *Journal of Engineering Science and Technology*. His research interests include thermo-fluids, renewable energy, biomechanical engineering, engineering education and academic leadership. He has numerous research publications, awards and honours. Mushtak is a Fellow of the Institution of Mechanical Engineers (UK), a member of the Global Engineering Deans Council and an Honorary Chair at the College of Engineering and Physical Sciences, University of Birmingham (UK).

Think Like an Engineer

Use Systematic Thinking to Solve Everyday Challenges &
Unlock the Inherent Values in Them

Dream Big. Be Different. Have Fun

by
Mushtak Al-Atabi
Author of "Shoot the Boss" and "Driving Performance"

Cover art by Chong Voon Lyn
Book typesetting and design by Mike Ooi

First Printing: 2014
Printed by MPH Group Printing (M) Sdn Bhd

ISBN 978-1500972288

Published by:
Mushtak Al-Atabi
No 1 Jalan Venna P5/2,
Precinct 5, 62200 Putrajaya,
Wilayah Persekutuan Putrajaya,
Malaysia

www.thinklikeanengineer.org

Acknowledgement

Writing this book was a very fulfilling journey. As it is true for all journeys, arriving to the destination is made possible by the help and generosity of many people. Here I wish to thank few individuals for their inspiration, help and guidance. First of all, I wish to thank my wife, who continues to be loving, patient and supportive despite my endless hours at work. I also wish to thank my kids, Sarmad, Furat and Ayham for their boundless curiosity, you guys are natural engineers. The work of Mike Ooi, my Masters student, in designing the book, typesetting it and taking most of the photos which was essential for getting my point across, and for this I am thankful. I am also indebted for Yvonne Lim, my PhD student; and Prof. Paul Peercy, Dean, College of Engineering, University of Wisconsin -Madison for proofreading the manuscript. Gratitude is also due to my students, both on campus and online, I have learnt a lot from all of you and I thank you for the successes we had, the failures we have learnt from, and most of all for sharing your stories. A number of my colleagues have contributed to the projects which I have described in this book, they have also supported many of my initiatives, no matter how crazy the initiatives may sound. I particularly wish to thank Edwin Chung, Satesh Namasivayam, Mohammad Hosseini, Chong Chien Hwa, Veena Doshi, Lim Chin Hong, and Nurhazwani Ismail. Last but not least, I wish to acknowledge Chong Voon Lyn for the beautiful design which graces the cover of this book.

Contents

Introduction		2
Chapter 1		4
Engineering		
1.1	What is Engineering?	5
1.2	CDIO Process	5
1.3	Civilisation, Engineering and Technology	10
1.4	Future of Humankind and the Grand Challenges	13
1.5	Society's Regulation of Engineering	17
Chapter 2		19
Brainology: The Biology, Psychology & Technology of the Brain		
2.1	The Brain	21
2.2.1	The Old Brain	24
2.2.2	The Middle Brain	24
2.2.3	The New Brain	24
2.2	The Brain, Learning and Mindset	26
2.3	How to Craft a Mindset	30
2.3.1	Brain Rewiring	30
2.3.2	Language and Thinking	31
2.3.3	CDIO Process as a Mental Model	34
2.4	Thinking and Reasoning	35
2.5	Critical Thinking	36
2.6	Knowledge and Memory	38
Chapter 3		42
Emotional Intelligence		
3.1	What is Intelligence?	43
3.2	Emotional Intelligence	45
3.2.1	Self Awareness	46
3.2.2	Self-Management	48
3.2.3	Social Awareness	53
3.2.4	Relationship Management	58

Chapter 4   67
Conceive
4.1   Preparing to Conceive   69
4.2   Ideation: The Art of Idea Generation   71
4.2.1   Brainstorming   71
4.2.2   Random Entry   74
4.2.3   Trimming   77
4.2.4   Blue Ocean Strategy   78
4.2.5   Trend Recognition   82
4.2.6   Biomimicry and Learning from Nature   89
4.3   Concept Evaluation and Selection   89

Chapter 5   93
Design
5.1   Function and Form   93
5.2   Design Process   94
5.2.1   System Architecture   95
5.2.2   Configuration Design   97
5.2.3   Integrated Design   98
5.2.4   Detailed Design   98
5.3   Design Optimisation and Trade Offs   99
5.4   Other Design Considerations   100

Chapter 6   104
Implement
6.1   Hardware Manufacturing Process   104
6.2   Software Implementing Process   105
6.3   Hardware Software Integration   106
6.4   Testing, Verification, Validation, and Certification   106

Chapter 7   110
Operate
7.1   Sustainable and Safe Operations   110
7.2   Operations Management   110
7.3   Training and Operations   111
7.4   The Birth of the Checklist   111
7.5   Preventive Maintenance   115
7.6   System Improvement and Evolution   115
7.7   End of Life Issues   115
7.8   CDIO Case Studies   115

7.8.1   Taylor's Racing Team   116
7.8.2   Women in Engineering   121
7.8.3   CDIO Beyond Engineering   124

Chapter 8   126
Ergonomics: Human Centred Design
8.1   Design for Ease of Use and Operation   127
8.1.1   Affordance, Visibility and Feedback   128
8.1.2   Constraints   130
8.1.3   Mapping   131
8.2   Anthropometric Measurements   132
8.3   Work Musculo-Skeletal Disorders (WMSDs)   133
8.4   Cognitive Ergonomics   133
8.4.1   Nudges   134
8.4.2   Framing   134
8.4.3   Anchoring   136

Chapter 9   137
Communication and Teamwork
9.1   Communication Strategies   137
9.2   Process Documentation   145
9.2.1   Logbooks   145
9.2.2   Minutes of Meetings   146
9.2.3   Technical Reports   146
9.2.4   Communication via Email   148
9.2.5   Operation Manuals   149
9.3   Teamwork   150
9.3.1   Selecting the Team Members   150
9.3.2   Core Competency   151
9.3.3   Organisation Chart   152
9.3.4   Team Evolution   153
9.3.5   Team Building Exercises   156
9.4   Connections and Networking   156

Chapter 10   159
Managing Projects for Success
10.1   Project Initiating   160
10.2   Project Planning   162
10.2.1   Cost and Resources Estimation   163
10.2.2   Time Estimation   163

10.2.3	Risk Management	168
10.3	Project Executing	169
10.4	Project Monitoring and Controlling	169
10.5	Project Closing	169
10.6	Project Stakeholders Management	170

Chapter 11 173
Entrepreneurship and Innovation

11.1	Philosophy of Entrepreneurship	174
11.2	Business Value	176
11.2.1	Customer Value (Value Proposition)	176
11.2.2	Shareholder Value	177
11.2.3	Employee Value	178
11.2.4	Suppliers and Partners Value	178
11.2.5	Society Value	178
11.3	Business Model (Entrepreneurial Ecosystem)	178
11.3.1	Customer	180
11.3.2	Channels	180
11.3.3	Revenue and Cost	180
11.3.4	Resources	181
11.3.5	Business Activities	181
11.3.6	Partnerships	181
11.3.7	Competitors	182
11.3.8	Business Environment	182
11.4	Business Plan	182
11.5	CDIO for the Market	184
11.5.1	Market Lifecycle of Products and Services	184
11.5.2	Market Evolution and Disruptive Innovation	186
11.5.3	Market Penetration	187
11.5.4	Customers Power in a Digital World	189
11.6	Lean Entrepreneurship	191
11.7	Funding Entrepreneurship	193

Chapter 12 198
Return on Failure

12.1	Learning and Failure	200
12.2	Failing Smart	203
12.3	Failure, Risk and Uncertainty	204
12.4	Education and Encouraging Failure	205

Chapter 13   209
Structured P Solving
13.1	Challenge Identification and Formulation	209
13.2	Root Cause Analysis	210
13.2.1	The 5 Whys	210
13.2.2	The 4 M Method	211
13.2.3	Fishbone Diagram	213
13.3	Estimation and Quantitative Analysis	214
13.4	Final Steps	215

Chapter 14   216
Engineering Holistic Education
14.1	Holistic Education	218
14.2	Learning Beyond the School	220
14.3	Mission Zero: A Vision for Higher Education	224
14.3.1	Zero Tuition Fees	224
14.3.2	Zero Impact on the Job Market	226
14.4	Engineering a Culture Change: Happiness Index	226

Epilogue   228
An Invitation

Foreword

Whether you are a professional engineer, an aspiring engineer, an enthusiast of engineering or someone interested in ideas, this book has something to offer. Whether you are interested in gaining an insight on how challenges are addressed, from the small, to the engineering Grand Challenges of our time, this book has something to offer. From energy, to the environment, to health, to aerospace, to construction, to information technology, this book takes you on a tour of engineering. But it goes much further than that; it considers how we conceive ideas, how the brain subconsciously imagines the products and the operation of engineering. The book highlights, in a new, innovative and readable way, how engineering is almost as depended upon emotional intelligence, as it is on the principles of mechanics.

This book is not only about "What is Engineering?" many books have attempted to answer this question, but goes much further and asks the question of "What makes Engineering?". It describes the melting pot of ingredients – the materials and inspiration from the natural world; the human spirit for challenge, change and the impetus to mould the world for our benefit; the human fascination for art, art form and creativity; our ability to think, abstractly, creatively and with a purpose; and finally our desire to influence the world around us.

In addressing these issues, Prof. Mushtak Al-Atabi takes you on an exciting tour that delves into the history of human inspiration, motivation and invention, the innate desire to solve challenges and the products that influence our daily lives and change the way we view world around us, now and into the future.

This book is an example, in its own right, of challenging thought, stimulating the mind of the reader. Whatever your background, and whatever your training, this book will make you think!

Prof. Nigel P. Weatherill, Vice-Chancellor and Chief Executive
Liverpool John Moores University

Preface

"At last, a book that supports science and engineering undergraduates to learn about understanding and developing their creative thinking and team skills to address the global challenges facing society. Engineers and scientists are often faced with finding and implementing solutions to pressing societal needs. This requires the simultaneous deployment of intellectual, technological and social skills. This book explores some of the things engineers spend insufficient time considering – the process of how to achieve a complex goal by working together.

For example, with the process of product manufacturing, and its associated complex supply chains, becoming a truly global activity a different attitude to education is needed. In the global mind-set paradigm there is still the same fundamental need for intellectual rigor but this needs to be applied within a multidisciplinary team context, perhaps with different cultural values. Linear thinking is unlikely to yield the correct outcomes in a sufficiently competitive timescale or cost envelope. Think Like and Engineer explores this creative paradigm.

Professor Mushtak Al-Atabi is well known for his inspirational teaching of engineering and entrepreneurship in classrooms and on-line. This book, written for undergraduate students, shares some of his thinking and passion. The text is illustrated with practical examples. The first few chapters allow students to see how the engineering brain needs to be 'wired' including developing an emotional intelligence in thought and action. In this context it introduces the processes of creating concepts to address engineering challenges, designing solutions and then implementing and operating the solution. This approach has become well known under the acronym CDIO in recent years. For me, a most important part of the book was the chapter discussing the dynamics in taking risks as an individual and as a team. In my experience this topic is rarely addressed and yet is so critical to understanding how more creative inventions can be turned into radical innovations. The book shows how engineering design groups can be assessed for their approach to risk-taking.

In summary an enjoyable and essential read for student scientists, engineers, and educators who want to impact society by thinking how they can develop a systematic approach to solving challenges."

Prof. Richard A Williams, Pro-Vice Chancellor, University of Birmingham

Introduction

I have been contemplating writing a textbook to accompany the project-based learning courses that I have been teaching for the past 15 years. As a matter of fact, the first drafts of the book I was thinking of are saved on floppy disks and I do not have the technology to read them anymore! This perhaps is a good thing as it gave me the opportunity to keep evolving what the book should be about and how it is pitched. I started with the idea of writing a book that outlines the skills and techniques needed by engineers and engineering students to run successful design projects. Supervising hundreds of engineering projects myself, I now can attest that skills and techniques for design are necessary, but not enough to make a good design engineer. What I experienced again and again is that successful design projects are often driven by a powerful dream that is backed by endless passion.

Working with student teams and sharing their moments of triumph as well as their moments of defeat, my perspective on design, engineering and project-based learning has profoundly changed. My current frame of mind is that the best way to prepare graduates for life is to create an environment in which they can find their passion and cultivate a growth mindset that views failure as a necessary prerequisite for success and learning. An environment where students can dream big, be different and have fun is the prerequisite for them to be able to develop self-awareness, positive attitudes and confidence that they can contribute to making the world a better place. Once this foundation is laid, provision of skills and techniques can go a long way to developing effective and sustainable designs.

Besides my academic and teaching duties, I also consulted for, and trained, executives from multi-national corporations. The clients I served ranged from technology firms to energy companies and banks. What I noticed was that the clients with non-engineering backgrounds often appreciate the engineering structure that I bring when I work with them to address the challenges they are working on. Some clients were even surprised with the way we train engineers and shared with me that this should be the way all education is conducted. This book is about how to train an engineering mindset in a project based learning environment. However, it is ultimately about human achievement and it can be easily used at any level including self-help. The aim here is to start transforming, or rather bringing back, higher education to have its goal to be the achievement of human potential, with obtaining employment and economic freedom as part of it, rather than having employment as the goal in and of itself.

While this book is primarily about engineering and how engineers think, address technical challenges and realise opportunities, the content in the book is written in such a way that it can be used by students and professionals in any field to develop a systematic approach to address the challenges in almost any field from marketing to medicine.

The book is based on the intuitive process of CDIO (Conceive, Design, Implement, Operate) which traces the lifecycle of a product, service or process. It also draws on the latest brain and thinking research, making it applicable in many human interactions. The CDIO process was initially developed as a framework to reform engineering education to enable the universities to produce "engineers that can engineer".

This book is aimed to be both a textbook for engineering and science students as well as a general read for anyone who is interested in engineering, system and critical thinking and engineering education. Engineering here is presented in the wider meaning of the word when it refers to cleverly and creatively orchestrating resources and components to provide value-adding solutions. To achieve this, the book features chapters dedicated to thinking and how the brain works, as well as emotional intelligence and practical techniques to cultivate positive attitudes, besides the systematic Conceive, Design, Implement and Operate chapters.

The ultimate aspiration of the book is to provide a practical and concise, yet holistic framework that can be used by anyone to solve challenges and add value while maintaining a balanced view of life. It also aspires to embrace the evolutionary nature of the creative process while providing the systematic scaffolding that ensures eventual convergence on the suitable solution.

Chapter 1
Engineering

*"Manufacturing is more than just putting parts together. It's coming up with ideas, testing principles and perfecting the **engineering**, as well as final assembly."*

Sir James Dyson, *Founder of Dyson Company*

*"The path to the CEO's office should not be through the CFO's office, and it should not be through the marketing department. It needs to be through **engineering** and design."*

Elon Musk, *CEO & CTO of Tesla*
CEO & Chief Product Architect of Tesla Motors

Engineering is old; as old as human civilisation itself. It was born out of the need of humans to modify their environment to provide their basic needs such as food, shelter and security. It has always been largely motivated by the human desire to have a better, easier and more comfortable life. Some of the ancient engineering marvels such as the wheel and the pyramids are still with us today and they will be around to amaze future generations for millennia to come. These marvels were achieved by talented individuals who were clever enough to manipulate the laws of nature to achieve incredible feats. While the ancient engineers did not go to universities and graduate with engineering degrees, they still needed to observe, study and hone their skills, to be able to achieve what they have achieved. Today, Science, Technology, Engineering and Mathematics (STEM) education has developed into an intentional and systematic field of study. STEM is considered, the world over, by policy makers, the industry and the community at large as being an extremely important bastion for economic prosperity, stability and security. While engineering thinking and mindset are often associated with the creation of technological solutions, clearly it can be used well beyond that. As engineering thinking and mindset are credited with seeing the non-obvious solutions and opportunities, it can be applied to a variety of world challenges in management, business, policy making and healthcare to yield high value. As a matter of fact, the Business Insider reported in 2013 that "33% of the S&P 500 CEOs' undergraduate degrees are in engineering, and only 11% are in business administration".

This book will try to outline and explain the engineering mindset and thinking techniques in a manner that is accessible and useful both to engineers and non-engineers so they can apply them in both engineering and non-engineering contexts.

1.1   What is Engineering?

Engineering is the professional discipline that is responsible for adding value through systematically applying the principles of science and mathematics to Conceive, Design, Implement and Operate (CDIO) products, services, technologies, systems and solutions that improve the quality of life. Using the CDIO process, engineers are able to use resources such as raw materials, energy and information to develop useful products supporting economic growth in a wide range of human activities.

Today there are numerous engineering disciplines to address the wide variety of industrial needs and economic applications. These include aeronautical, civil, chemical, computer, electrical, electronic, mechanical, and petroleum engineering. Emerging engineering disciplines, such as biomedical and mechatronic engineering; are developed in response to the increasing complexity and multidisciplinary nature of the world we live in today.

1.2   CDIO Process

It is said that technology is the manifestation of our dreams. This describes engineering very accurately as it takes our dreams and makes them a reality through the Conceive, Design, Implement, Operate (CDIO) process. The CDIO process reflects the life cycle of technology-based products and services, and the best way to demonstrate this process is through the journey of a product. Since I like airplanes, I shall choose the A380 as an example to demonstrate the CDIO process.

The A380 is the world's largest passenger plane. Manufactured by Airbus, the plane has a wingspan of 80 meters and is able to reach a speed of over 1000 km/hour. For the A380 to be able to serve its purpose, countless engineers, designers, technicians, managers, financial experts and lawyers needed to work on Conceiving, Designing, Implementing and Operating it.

The Airbus A380 (Source: Wiki Commons)

Everything that is intentionally made by a person starts as a thought in the brain. Hence the first part of the process of making anything is to Conceive it. I presume that the idea to make the world's largest passenger plane was toyed with by the Airbus technical and management team and many discussions took place before approving the idea. The discussions would have taken into account the economic outlook of the market, and the customer needs and desires, as well as the technological possibilities and limitations. In the context of engineering a product, service or a system this stage is called Conceiving. We can Conceive an entirely novel product, such as a new fuel or a new method to communicate, a small part of a bigger system, or we can Conceive a new version of an existing product, such as a plane that will have the same basic components like any other similar plane. So the Conceiving stage of an A380 will result in an "idea" of building a large plane that will serve a certain segment of the market. This stage requires a thorough understanding of the business, regulatory, and technological context; and can involve sophisticated teamwork and various focus groups to enable the complex decision making required to venture into such a long term commitment to make a passenger aeroplane.

Conceiving the plane as a super-system will set into action numerous processes to CDIO subsystems necessary to make the plane a reality; however, we shall focus on the plane itself to demonstrate the CDIO process. Subsystems include airframe, jet engine, avionics, etc.

With the Conceiving stage completed, the Design stage can start. Here engineers will be making a series of complex technological calculations, tests, decisions, and compromises. There will be many conflicting requirements to consider, including safety, technical, economical, and legal requirements. Calculations and experiments will aid decisions on considerations including how big the plane will be and how many engines it will have. A key feature of the Design stage is to anticipate where and when any part of the plane may fail and to try to prevent this. Predicting, pre-empting and preventing failure are important skills that engineers need to hone to be able to build reliable products that are safe to use time and time again. Interestingly, in order to prevent future catastrophic failure, engineers need to ensure that they push product parts and components to the limit and make them fail. This way they will understand the limits of safe operation of the systems they are making. To achieve this, engineers make airplane models that they push to the limit in wind tunnels, and computer models that they simulate in computers using specialised software. Eventually the Design stage will yield detailed engineering drawings that outline the sizes, materials and specifications of all the parts of the plane.

Notice that after completing the Design stage, the plane is still far from being a reality. It has moved from being a thought or an idea to being a series of drawings and lists of materials on papers and computers. Our plane is now ready to come to life - to be Implemented. The Implementation stage is when engineers take their designs and transform them into real products. In the case of the A380, engineers need to decide how, where and when each part of the plane will be made and tested and how it will be integrated into the super-system (the plane). While the final assembly and testing of the plane takes place at Airbus factories in Toulouse, France, different parts are manufactured all around the world. For example, the jet engines of the plane are made by Rolls Royce in the UK; the tyres are made by other tyre manufacturers, and so on. Fast forward and we have a plane ready for delivery, complete with custom paint and interiors as requested by the customer. But the process does not end here.

Engineers still need to think of how the plane will be Operated safely after its delivery, this is done through the preparation of service and maintenance standards, as well as training and licensing programmes for the staff that will maintain the plane while it is being operated by the airline. Modern engineering is pushing the limits through thinking of how the airplane can be modified and upgraded while being operational, as well as having plans on how its parts will be recycled when it is eventually retired.

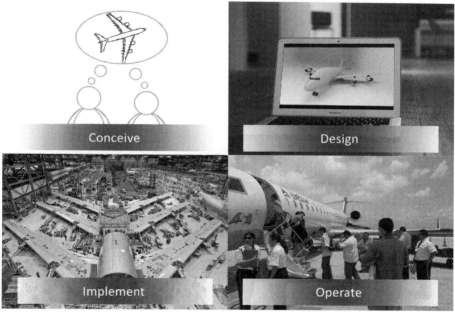

CDIO of an Aeroplane (Source: Wiki Commons)

I hope that this quick account has shed some light on the CDIO (Conceive-Design-Implement-Operate) process. It is useful to mention here that the account provided above is a macro-CDIO, within it each and every component of the plane has been through its own CDIO process as well. The process is logical and can be used by everyone to bring new ideas to life. It can be used to make a toaster, a car or a space shuttle.

The CDIO process can also be deployed when developing non-engineering products. Think of a new surgical procedure or a new insurance product. Both are Conceived, Designed, Implemented and Operated. Part of the agenda of this book is to "sell" the CDIO process as a thinking system that can be used well beyond traditional engineering realms.

When working on each stage of the CDIO process, engineers ask themselves the following questions in relation to the product they are making and how are they making it.

1. Is what I am making desirable? Does the customer need it, want it and like it? Does it solve a challenge or satisfy a need?

2. Is it economically viable? Will the customer be willing to pay for it? Is there a way to make it cheaper, more productive and add more value?

3. Is it feasible? Is the technology to make it available?

4. Is it ethical and legal? Does making or selling it infringe on any law or intellectual property (IP)?

5. How can I make it safe for both those who make it and those who use it?

6. How can I make it easy to make and use?

7. How can I reduce its impact on the environment even when it is no longer in use?

This book has a dedicate chapter for each part of the CDIO process. Each stage will be explained in further detail, and the tools to help us better Conceive, Design, Implement and Operate will be shared.

It is worth iterating here that the CDIO framework, at its essence, is a systematic and critical thinking framework. Practicing the CDIO process will build mental habits that can be used in solving everyday challenges and unlocking the inherent opportunities stored in these challenges. It provides a balanced way that embraces the iterative and evolutionary nature of the creative process while maintaining a systematic framework that ensures the fulfilment of the set objectives. In short, whether the challenge being addressed is in business, marketing, technology, entrepreneurship, academia or even in one's personal life, CDIO process can provide an effective way to yield a solution.

1.3   Civilisation, Engineering and Technology

Since 2.6 million years ago, when humans began altering their environment using stone tools, major changes that have occurred have always been associated with one form of technology or another. Technological revolutions are marked by rendering a scarce resource (material, energy, information, knowledge) more widely available. This often is accompanied by far-reaching economic, political and social consequences. Technological revolutions bring about radical changes so that, literally, life is never the same again after their introduction. Throughout history, a number of technological revolutions shaped the world that we live in. A list of these revolutions is given below.

1. **Putting fire under control (800,000 BC)**
   Controlling fire was a great technological feat. It gave humans access to energy when it's most needed, i.e., for warmth and cooking food; which in turn enabled humans to extract more calories from the food they eat, hence improving their survival chances.

2. **Agricultural Revolution (8000 BC)**
   Unfolded in Mesopotamia, modern day Iraq, mastering agriculture was a real technological triumph. It was the result of putting huge amounts of knowledge and observation into a practical use. Agriculture permitted humans to get more out of the land they lived on and aided the formation of the first cities. The same piece of land, when used for agriculture, can support far more people compared to the same area when used for hunting and gathering. Living together in cities had a huge impact on the way civilisation evolved as it enabled the concentration of cognitive capital that led to the accumulation, growth and recording of knowledge, which was the basis for all future human development.

3. **Invention of the Wheel (3500 BC)**
   Although it is very difficult to imagine our world without wheels, there was a time when there were no wheels to go around. Transportation was very difficult and ineffective. Archaeological evidence supports that the wheel was first invented in Mesopotamia as well. As it is true for other technological evolutions, the invention of the wheel set humanity in motion in many ways. This paved the way to many future human developments and revolutions.

4. **Scientific Revolution (1534-1700 AD)**

    Although the idea of the printing press appeared in different parts of the world, the first use of a printing press to print books is attributed to Johannes Gutenberg, who used the printing press in 1450 AD to print copies of the Bible. Coupled with other innovations, this technology helped unlock the resource of information as more people could have access to books, which used to be very scarce and expensive. Continuous recording, organising and refining of information into knowledge and spreading it led to scientific reasoning and the questioning of the religious views on how the universe works. The publication of *De revolutionibus orbium coelestium* (On the Revolutions of the Heavenly Spheres) by Nicolaus Copernicus and *De humani corporis fabrica* (On the Fabric of the Human body) by Andreas Vesalius are often cited as the marking of the beginning of the scientific revolution and the age of reason. This tradition of scientific reasoning opened the doors for millions of people to question, understand, and build on the knowledge developed by others.

5. **Industrial Revolution (1712 AD)**

    The development of the steam engine led to the dawn of the industrial revolution. Energy stored in coal can now be unlocked and used to power trains and machines that decreased the cost of production and brought about the era of mass production and consumerism with all of its economic, political, social and cultural implications. As is true for all true technological revolutions, the industrial revolution had far-reaching political and economic consequences.

6. **Digital Revolution (1950 AD)**

    Programmable machines are directly connected to the development of large knitting machines that used punched cards to automatically operate machines to produce pre-designed patterns. The first programmable computer was designed in 1936, but the 1950s are generally accepted as the beginning of the digital revolution era that we still live in today where computers became so ubiquitous that we do not even feel their presence while they take care of our cars, fridges and washing machines.

7. **Information & Communication Revolution**

As the computers became more powerful, cheap and easy to use, they ushered in the digital era which is synonymous with the Internet and the World Wide Web. The information and communication revolution changed the way we work, study, trade, communicate and entertain ourselves. This created a virtual world that is now so vast, it competes with the real world, and there are researchers who predict that the digital elements of our lives will end up redefining what it is meant by being a human.

8. **Knowledge Revolution**

The oversupply of huge amounts of information free of charge and on demand is changing the world beyond recognition. This is especially true in areas of education where education systems are evolving to reflect the reality that education is about constructing knowledge rather than just remembering facts. The knowledge revolution is correlated with the rise of knowledge economy where information is constructed and organised into knowledge that can be utilised to create economic value. Knowledge management is also allowing us to gradually use machines to perform tasks that need complex decision making.

9. **Green Revolution**

The colour of the 21st Century is green. As sustainability takes centre stage, the green economy and green development are moving towards the mainstream of the political, cultural, technological and educational debate. Being sustainable is gradually becoming a requirement for all our activities, and this is having a major impact and changing the way we do many things including the way we travel, manufacture goods and produce our energy and food.

The above is a brief account of the story of humankind from a technological perspective. It is clear that technology has played, and will continue to play, key roles in advancing the human cause. This needs to be balanced with a holistic personal and professional development to ensure that we have the wisdom to use the power unlocked by technology in a constructive and sustainable manner.

1.4   Future of Humankind and the Grand Challenges

Now, more than any other time in the history of humankind, engineers are required to provide solutions to ensure the survival of the human race. The National Academy for Engineering (USA) identified 14 Grand Challenges that we need to address in the 21st Century to have a chance to make it into the 22nd Century. These Grand Challenges are listed below and they can be used to guide not only engineers as they pursue their careers, but also business leaders, policy makers and educators as they go about making the complex decisions needed on daily basis. Awareness of these Grand Challenges can provide us with a strategic horizon when making decisions related to career, business, policy and education.

Energy and Environment

1.   **Make solar energy economical**
    The sun represents a wonderful source of energy that cannot be matched by anything humans have made thus far. Although only a fraction of the solar energy arrives to the earth's surface, it is sufficient to provide 10,000 times the energy needed to power all the commercial activities that the planet needs. But there is a catch, solar technology is not economical yet! Current photovoltaic cells are only 10-20% efficient, and the cost of manufacturing them are still relatively high. Another challenge that is associated with solar energy is the inadequacy of energy storage media to capture the solar energy and store it and transport it to be used when and where it is most needed. Clearly making solar energy economical is a pressing need and this will continue to captivate and motivate scientists and engineers in this century.

2.   **Provide energy from fusion**
    Fusion is how the sun makes its energy. Single proton nuclei of two hydrogen isotopes are fused together to produce the heavier nucleus of helium and a neutron. In the process some mass is converted into energy following Einstein's equation $E=mc^2$. Although fusion has been demonstrated on a small scale, the engineering challenge is to build a reactor that can withstand the huge pressure and temperature associated with the fusion reaction. As hydrogen is the most abundant material in the universe, you can imagine how much energy we can have when the technology surrounding fusion is mastered.

3. **Develop methods for carbon sequestration**

Commercial and industrial human activities are often associated with emission of carbon dioxide, among other gasses. This is now becoming a challenge as carbon dioxide is contributing to climate change. Carbon sequestration involves storing carbon dioxide away from the atmosphere. There are a numbers of ways to achieve this, including underground and underwater storage, and storage in depleted oil and gas reservoirs. The technology is still expensive; however, this area is a hotbed for innovation where researchers are looking for solutions that can convert carbon dioxide from waste to wealth, i.e., converting it into something useful rather than just storing it away.

4. **Manage the nitrogen cycle**

Nitrogen represents around 79% of the atmospheric air, and it represents a very important substance for life as it is an essential component of amino acids which in turn are building blocks for proteins. Living organisms, including humans cannot utilise atmospheric nitrogen directly, nitrogen needs to be combined with carbon before it can be absorbed by living organisms. This important stage is done by the plants we eat with the help of some microorganisms. To complete the nitrogen cycle, some organisms use nitrogen nutrients as a source of energy and return nitrogen molecules back to the atmosphere. With more and more food production, humans are putting more nitrogen-based fertilisers into the environment, and these are not being converted back into free nitrogen. These nitrogen-based compounds are contributing to the environmental pollution. Engineers need to devise innovative ways to improve the efficiency of various nitrogen related activities to ensure that nitrogen-based fertilisers are minimally and effectively used and any unused fertilisers are converted back into free nitrogen.

5. **Provide access to clean water**

It is heart wrenching that 1 out of every 6 people living today does not have access to clean running water. Access to clean running water and adequate sanitation is an important requirement to leading a healthy life. Although 70% of the earth is covered with water, most of this water is not suitable for human consumption, either because it is polluted or is salty seawater. Engineers can play a key role in providing access to clean water for both drinking and other activities, such as agriculture, through the development of technologies that can help purify polluted water and prevent more water from becoming unsuitable for human use.

Health

6. **Advance health informatics**

Information technology has the potential to improve healthcare and reduce its cost. This can be done through digitising existing medical records for individuals and groups, and producing a robust health information system that enables health professionals to detect, track and respond to health emergencies and pandemics. This area will continue to be a fertile field for innovation and value creation, especially as sizable proportions of national budgets are dedicated to healthcare.

7. **Engineer better medicines**

Although every person is unique and different from others, the medicine prescribed by our doctors to address a given disease is often the same. The development of equipment for fast genetic profiling and organism-specific antibiotics can help revolutionise personalised medicine, and biomedical engineers are expected to play a key role in this. Engineers can also contribute towards the creation of medicines and treatments that target only the diseased tissue, reducing side effects and speeding up recovery. One of the promising areas of development is the work engineers are currently doing with surgeons on brain implants that enable disabled individuals regain control over their body parts.

Security

8. **Prevent nuclear terror**

Engineers have the responsibility to protect society against the potential threat of terrorist attacks using radioactive materials. This is mainly done through securing radioactive waste and preventing unauthorised access to it as well as developing technologies to detect and respond to any attack.

9. **Secure cyberspace**

Cyberspace is a very important domain in which we are spending an increasing amount of time and performing more and more activities, especially trade and communication. With highly sensitive national and personal information, as well as billions of dollars in transactions going through the Internet, cyberspace is becoming a field where criminals and terrorists operate. Computer and software engineers need to develop solutions that make online interactions safer for everybody.

10. **Restore and improve urban infrastructure**
Infrastructure systems are the foundation on which our civilisation is built. They include water, sewer, roads, power, and telecommunication systems. Engineers are facing the huge challenge of modernising the infrastructure systems for cities to enable them to continue to support the increasing human population.

Learning and Computation

11. **Reverse engineer the brain**
The human brain represents one of the final frontiers of human discovery. It is highly effective and efficient, and no computer can match the capabilities of the brain. Learning how the brain thinks, learns, and stores and retrieves memories and information, will have tremendous impact on our capability to build powerful and efficient computers.

12. **Enhance virtual reality**
Although virtual reality is often associated with computer games and entertainment, enhancing virtual reality can have a great positive impact on the more practical aspects of our lives. It can help engineers design new products and allow users to test them before making them. Surgeons can perform virtual operations before cutting into real people.

13. **Advance personalised learning**
We all learn differently. Technologies are increasingly available to allow personalisation of the learning experience, taking into account the different learning styles, paces and preferences of different learners. Currently web based courses and Massive Open Online Courses (MOOCs) represent the beginning of this wave, and the future carries more exciting possibilities, including real time monitoring of brain activities of learners to ensure the most effective and rewarding learning experience.

14. **Engineer the tools of scientific discovery**
Engineers will continue to Conceive, Design, Implement and Operate tools and equipment that will enable scientific discovery, allowing us to

advance our knowledge and understanding of the universe. From telescopes to explore the far corners of the universe, to microscopes that allow us to study the smallest molecules; and from rockets to explore outer space, to remotely operated vehicles to probe the depths of the ocean, engineers will continue to play key roles of building the tools of scientific discovery.

1.5 Society's Regulation of Engineering

Engineering is an important profession. Engineers often make technical decisions that can impact the safety of the public as well as the allocation of huge amounts of money. Just like other professional practices such as medicine, accountancy and law, the practice of engineering is legally regulated by professional bodies. These engineering professional bodies perform a number of very important duties, which include licensing engineers to practice through regulation of their membership; accreditation of engineering degree programmes offered by universities and ensuring that these programmes meet the necessary professional requirements; and approval of the technical standards used for designing different engineering components.

1. Engineers think and add value using the CDIO process

   • Conceive
   • Design
   • Implement
   • Operate

2. The Grand Challenges for Engineering in the 21st Century are:

   • Make solar energy economical
   • Provide energy from fusion
   • Develop carbon sequestration methods
   • Manage the nitrogen cycle
   • Provide access to clean water
   • Restore and improve urban infrastructure
   • Advance health informatics
   • Engineer better medicines
   • Reverse-engineer the brain
   • Prevent nuclear terror
   • Secure cyberspace
   • Enhance virtual reality
   • Advance personalized learning
   • Engineer the tools of scientific discovery

Chapter 1
Engineering

Chapter 2
Brainology: The Biology, Psychology & Technology of the Brain

"The chief function of the body is to carry the brain around."

Thomas A. Edison

"There is nothing either good or bad but thinking makes it so."

William Shakespeare

"My greatest challenge has been to change the mindset of people. Mindsets play strange tricks on us. We see things the way our minds have instructed our eyes to see."

Muhammad Yunus, *Nobel Laureate*

Have you ever gazed at the moon or the clouds and happened to see faces or other patterns? If you have experienced this, do not despair; you are not alone! The human brain is hardwired to make sense of the world, and recognising patterns through comparing what it perceives to the patterns stored in it. The brain takes in stimuli from the environment and creates the perceived reality using the mental models available in the mind. The brain responds to stimuli, which can be either emotional or physical. Physical stimuli include visual, auditory, tactile, taste or smell. All these stimuli are converted into electro-chemical signals that the brain can process by comparing them to stored mental models. These mental models represent thinking patterns or templates in the mind that our brains use to understand the world and others around us, and how we relate to the bigger scheme of things. Bombarded with continuous stream of stimuli, mental models help impose order on the chaos of the world and select the stimuli that are most relevant and important.

Mental models are simply the way we perceive the world. We are born with some mental models hardwired in our brains, while other mental models develop and get hardwired through learning, practice, experiences, attitudes and interactions with our environment and other individuals around us. Examples of mental models that are hardwired into our brain upon birth are the thinking templates which enable us to recognise and avoid danger and those which enable us to quickly recognise faces. While very useful to enable quick thinking and making the brain the formidable thinking machine that it is, mental models may come at a price. Sometimes we may unconsciously perceive danger in a non-dangerous situation or see faces in random patterns such as cloud formations or on the surface of the moon. Mental models are continuously acquired and hardwired as we go through life. They give us our mental habits and attitudes and ultimately form our mindsets. A mindset is a very important individual trait. It represents our character and governs how we generally perceive life. We may have a happy, positive and optimistic mindset or a negative and pessimistic mindset. Optimistic mindsets are often geared towards growth, enabling those who adopt them to continuously learn, transform and recognise opportunities.

As mentioned earlier, everything that is man-made starts as an idea in the brain, hence it is appropriate to dedicate a short chapter to this wonderful organ. It is also fitting to start with a caveat that we know very little about the brain and how it operates. Interestingly enough, reverse engineering the brain is one of the Grand Challenges for Engineering identified by the National Academy of Engineering in the United States. In this chapter, I will try to summarise the available knowledge about brain biology (hardware), psychology and thinking (software), and propose the CDIO process as the technology or thinking technique that allows us to effectively use our brains.

This knowledge of the biology, psychology and technology can have a profound impact on developing effective thinking skills that can result in high performance in study, work and life. In her book 'Mindset: The New Psychology of Success', Carol Dweck shared that when students are educated about the brain and how it works they learn more effectively. She cited an experiment where, at the beginning of the school term, students were randomly divided into two groups. One group was given a two-hour workshop about the brain and how it works, while the other group was not given this additional workshop. The results were amazing! The students who attended the workshop learned better and achieved better end-of-term results compared to those who did not attend the workshop. It is worth mentioning here that the teachers, who taught the students throughout the semester, were not aware of brain workshop, just to ensure that there was no bias in the treatment.

I personally make it a point to start my courses, both with my engineering students and corporate training clients, with sections on how the brain works. The hypothesis is that all learning takes place in the brain and understanding the brain a little bit better will help the learner select and adopt suitable learning techniques and strategies.

2.1   The brain

Biologically speaking, the basic building blocks of the brain tissue are neurons. These neurons are responsible for transferring electro-chemical signals to and from each other. These electro-chemical activities, in turn, are responsible for the brain functions and the emergence of the mind. Neurological research has made many advances in recent decades, backed by technological wonders such as FMRI (Functional Magnetic Resonance Imaging) machines, where the brain of an individual is scanned while that individual is performing a certain mental task. Knowing which part of the brain is engaged when each type of mental task is performed, we now can map different areas of the brain to the variety of activities that we perform and roughly know which part of the brain is activated when we are involved with a certain mental activity.

The human brain represents a very flexible system. It has a huge growth and learning potential and it continues to change throughout the lifetime of an individual, responding to different challenges and activities and continuously forming new connections between its different neurons. Neurons are responsible for transferring electro-chemical signals within the nervous system, enabling the emergence of the mind and the computing power of the brain to manifest itself as thoughts, decisions, memories, emotions and physical activities. The flexibility of the brain is an attribute of the possibility of creating billions of combinations of connections between brain neurons as we learn new things and have repeated experiences. It is now well established that we can "sculpt" our brains to be happy, positive and successful through intentional actions as well as mental and emotional exercises.

The Amazing Human Brain

One way to imagine the neurons in the brain is to imagine a bunch of electrical wires. Some of these wires have insulations but a considerable number of them are not well insulated. Because of this, neurons leak electrical signals to neighbouring neurons as the signal passes through, weakening them. That is why when we learn a new skill, it feels awkward in the beginning, but as we keep on repeating and practicing, a substance called myelin starts to form around the pathway of the electrical signal as it travels. Myelin is an electrical insulator and as more layers of myelin are wrapped around the neuron pathway, an equivalent of a signal highway is created, where very high speed neuron firing occurs. This is the reason why we get better at doing things as we practice them, and why it is difficult to change habits. In neuroscience, the saying goes like this "neurons that fire together, wire together."

Myelin formation starts at the 14th week of the embryonic development and continues throughout life. The growth of myelin in embryonic stage ensures the "programming" of the basic capabilities that an infant will need upon birth. Compared to babies of other animals, human babies are born more vulnerable and with little capabilities; a baby gazelle, for example, is born able to walk while a human baby needs a year before it can walk. This is an indication that the neuro-circuitry for walking is yet to be myelinated in humans upon birth. Although this may seem as a disadvantage at first, in actual fact the lack of myelination at birth is a powerful advantage. It gives the human brain unlimited potential as we are able to programme our brains through inducing the formation of myelin through neuron firing. This makes the human brain highly adaptable and enables individuals to master a highly diversified number of skills and professions.

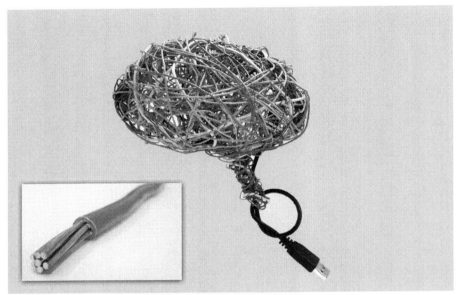

Myelin is Like the Insulation around Electric Wires

So whether learning how to play golf, how to play the piano, or how to drive a car, what we are doing is building myelin around the pathways of the electrical signals needed to control the muscles responsible for performing these tasks. Myelination results in better, faster and more targeted neuron firing sequences, and hence mastery of the skill. Interestingly, the same process works for intentional and non-intentional actions, so if we keep on repeating a routine long enough, myelination will lead to the wiring of our habits and mental models.

If we take a brain cross section along the body's plane of symmetry, three distinctive regions of the brain are identified. These regions are named the old brain, the middle brain and the new brain. If you imagine the brain to be a cone of ice-cream, the first scoop will represent the old brain; the second scoop will represent the middle brain while the new brain will be represented by the third scoop. Although different areas of the brains can be categorised in other more detailed ways, this categorisation of the brain is more useful in the context of this book. It is useful here to mention that thinking is a very complex process and often even the simplest decision involves multiple areas of the brain working together. Now let us examine the main features of each brain.

2.1.1 The Old Brain

It is also called the reptilian brain because it is very similar to the brain of the reptiles. In this part of the brain, the basic instincts are hardwired. The old brain is rather automatic and it gets activated in the event of danger and life-threatening conditions. We have very little control over this area of the brain. The old brain is responsible for the split second decisions we make when a car is about to hit us, or if we suspect that a venomous snake is about to attack us. So the old brain is all about survival; it gets really activated when we feel danger and it quickly responds to negative stimuli. If we were to briefly describe it, we could say it is fast and selfish, because it acts quickly and only in self-interest.

2.1.2 The Middle Brain

This part of the brain is where emotions reside and are managed. It is responsible for processing how we feel and it is not capable of processing language. This is may be the reason why we often find it difficult to express our feelings and emotions in words. Contrary to the common belief, humans make many decisions based on emotional rather than rational reasons. From a computing point of view, this simplifies and quickens the process of decision-making. This has huge implications for politicians, marketers and educators who would like to influence the behaviour of others. We vote, buy and learn when we "like" what we see.

2.1.3 The New Brain

This part of the brain is mainly responsible for high level reasoning such as mathematics and language. It is very analytical and slows the process of decision making. It is the rational brain.

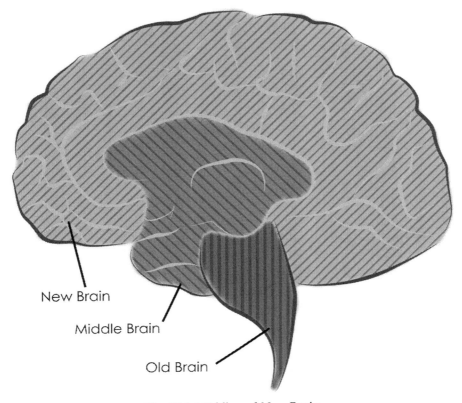

New Brain

Middle Brain

Old Brain

The Old, Middle and New Brain.

As mentioned earlier, contrary to the common belief, the emotional brain plays a key role in the decisions that we make. This was illustrated by the case of a patient known in the neuroscience literature as Elliot. Elliot had a surgery to remove a small tumour from his brain. His doctors were very pleased to note that the surgery did not impact Elliot's cognitive capacities or speech. However, strange things started happening to Elliot after the operation. He started to endlessly debate each and every detail in his life. Decisions such as which shirt to wear or which way to take to work took Elliot hours to ponder. Even simple things like whether to use a black or blue pen became stressful decisions, and Elliot weighed all their pros and cons before settling on the colour to use. Elliot was eventually sacked by his employer and his wife divorced him. Closer examination by neuroscientists revealed that although Elliot's new brain was doing a good job at the rational side of decision making, the surgery somehow disconnected Elliot from his emotional brain. Losing that "gut feeling", or "like or dislike" feeling that we all have when faced with everyday life decisions, rendered Elliot incapable of making any decision. In order to lead happy and successful lives, it is essential that all parts of our brain work together in harmony. It is also important to understand the limitations of our thinking so that we can improve its quality.

The above quick and brief description will prove useful as we explore areas such as emotional intelligence, communication, teamwork, human interaction and ergonomic designs later in this book. It is useful to reiterate here that the brain works like an orchestra with all parts of it contributing to the thinking symphony. So the three regions of the brain are always collaborating to respond to the continuous stream of stimuli that we call life.

2.2   The Brain, Learning and Mindset

As discussed earlier, the brain works through generating mental models that impose order on chaos and deduce patterns to compare with previous experiences, impacting the way we perceive ourselves, others and interpret the world around us. Different people have different mental models, and that is why we perceive the same events differently. For example, if two individuals are learning how to play a violin and both of them struggle equally with their learning, that struggle is nothing but an event that is happening to both of them as they push the limits of what they are capable of. Now each individual will use his or her mental models to interpret this event (the struggle). One mental model may be "playing the violin needs more practice; if I persevere long enough, I will be able to play beautiful musical pieces in due course". Another mental model may be "this is difficult for me. I do not have the talent for music; I will never be able to learn how to play the violin." It is clear how each mental model will drive each individual on a totally different path. The collection of mental models that we myelinate and cultivate result in our overall mindset and attitudes that drive how we perceive the world, our role in that world, and how we respond to the challenges we encounter. When faced with a new situation, one individual may see adversity, while another may see opportunity, depending on the mental model that they adopt.

Entrepreneurship, innovation and creativity are simply mental models that are adopted by individuals. Entrepreneurs, innovative, creative and highly motivated individuals cultivate the ability to generate positive mental models that are crafted to see opportunities in any new situation (just like seeing a face on the moon). Being able to see the opportunity even in adversity is literally the ability to generate positive and flexible mental models to see the world through. In his book 'Outliers', Malcolm Gladwell studied a number of high performing individuals including Bill Gates and Tiger Woods. He concluded that to achieve mastery level in a major completive discipline such as sport, art, science or business, individuals need to dedicate 10,000 hours of focused practice and training. This level of commitment and consistency is definitely an indication and precursor of the growth mindset. Whenever we see a master performing, we are tempted to think that the master is talented and lucky to be able to perform at this level. The fact of the matter is that mastery is always a product of a mindset that sees the struggle in new challenges as a stepping-stone towards progress and high performance. This is true for golf, chess, physics or music. All masters practice their craft religiously, building and strengthening myelin layers over the neuro circuits that they have cultivated over repeated and conscious practice.

To be successful in life and work, and to successfully Conceive, Design, Implement and Operate value-adding products, systems, and experiences, the ability to develop positive mental models is paramount. Developing these positive mental models requires committing the time and effort to practice and learn necessitating many sacrifices on the way. Decades of research in human motivation indicate that the main source of it is our search for meaning. We are positive, motivated and on the top of our game when we work on something that is meaningful to us. To achieve this, it is important to continue training and rewiring our brains to develop the muscles of positive thinking that are necessary to view challenges as opportunities so that we can, through them, create a better world.

There is enough evidence that we can exert control over what habits and mental models we form through repetition, reflection and intentional thinking as well as the use of experienced coaches and mentors. These intentional mental activities cause myelin growth and strengthen the neuron connections, literally hardwiring the mental models we are inculcating. Through this we are able to nurture positive mental models that will support growth, development and learning. The collections of mental models that we maintain will, literally, create our world and form our mindset.

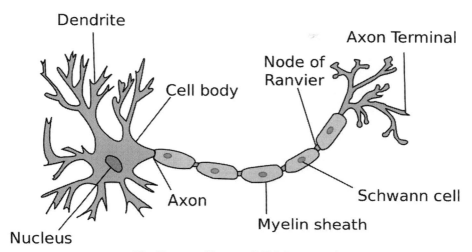

The Neuron. (Source: Wiki Commons)

Nurturing a positive growth-inclined mindset is very necessary for success both professionally and in life at large. Individuals who continuously develop their mindset are lifelong learners who will be able to fulfil their full potential and play a key role in helping others achieve their full potential as well. In order to be effective individuals who can develop people, products and systems that make life worth living and support the increasing population of our planet, it is important that we develop a positive attitude and see the world as full of opportunities and cultivate mental habits to construct mental models that enable us to spot those opportunities and create value out of them. Here we need to reiterate that our perception of reality is extremely important. Strictly speaking, there is no absolute reality, and that is why two individuals facing the same situation may end up seeing adversity or opportunity depending on their mental model.

Studying the biographies of successful individuals, who managed to have a lasting impact on those around them, seems to suggest a formula or a process for success. This formula seems to suggest that success happens when we work on something that we are interested in and willing to commit our time and effort towards. Interest and commitment, when coupled with the right practicing technique, coaching and feedback for improvement, will result in achieving mastery and success. These elements are discussed below.

Interest

The journey to success starts with interest, successful people often pursue goals that they are interested in and are in alignment with their life purpose. Simon Sinek calls this "starting with why". In his book carrying the same title, he cites many examples of organisations and individuals from Martin Luther King to Apple Inc. where successful and inspirational leaders are very clear of the "why", the reason for their existence, and use that to communicate and inspire. It is important for individuals and organisations that seek success, happiness and fulfilment to articulate their interests and life purposes as this is the key to sustaining them throughout the ups and downs of the life journey. Writing down a "Vision" statement, for both individuals and organisations, is a worthwhile effort in this direction. If done right, it can bring a lot of clarity to answer the question "what is the purpose of our existence?"

Commitment

Pursuing what interests us is an essential starting point but it is not sufficient for success. We need to be committed to allocate the time and effort necessary to build the success mindset. Willingness to push the limit and operate outside our comfort zone is a prerequisite for honing new skills and attitudes leading towards success. Interestingly, that willingness to explore uncharted territories is also the reason why we will inevitably fail before we achieve mastery. This "failure" is a necessary feedback that we should use to know what to change and where to improve in order to fare better next time. The way we respond to failure is a very good indication of the depth of our interest in what we are pursuing and it represents a reflection of our mindsets as well. An interested committed mindset will perceive failure as getting us a step closer to success through uncovering shortcomings and things to change.

It is necessary here to mention that the definition of failure in this section, and most of the book, is the inability to achieve the desired results when we push the limit of what we are capable of or what the current paradigms allow. This failure often yields very positive learning that can be used to further improve performance in the future.

Practicing Technique, Coaching and Feedback

Techniques are mental frameworks that we use to improve our performance while pursuing our developmental goals. Often these techniques are used under the tutelage of a coach, teacher, mentor or a master of some sort. Techniques refer to the way sportswomen and sportsmen train when they prepare for the Olympics, for example. The presence of a coach is often very important. I taught myself swimming at the age of 27. I had the interest and I was truly committed. I went to the swimming pool every day and watched how other people swim. Eventually, I was able to swim, but my technique is not good at all. I attribute this to the fact that I did not get someone to show me how to do it right and provide me with timely feedback so that I can correct and alter my technique and build myelin around the right brain circuitry.

The practice that leads to success takes place with full awareness on the part of the trainee while a selected limit is being pushed. For example when a runner is trying to break her own record, she may try a new rhythm of running or breathing, as she does this new breathing, her performance may take a dip, she needs to be aware of what is happening in her body and mind, and continue practicing fine tuning and perfecting the skill. When a limit is pushed, any failure is a feedback or symptom that indicates that further fine tuning is necessary, so failure should be expected, accepted and even welcomed as a sign that we are on the path to enhance and improve the skill.

2.3   How to Craft a Mindset

It is clear now that formation of mental models is a process that happens over time as myelin grows along the neuron pathways. We have a choice of allowing this process to take place unconsciously or to take charge of it and literally use myelin to craft our desired mindset. This section proposes three main techniques to growing a positive mindset, brain rewiring, language and thinking, and the CDIO process as a mental model.

2.3.1   Brain Rewiring

We mentioned earlier that mental models to recognise danger exist much earlier in the evolution of the brain as they can be a necessary prerequisite for survival. However, mental models that recognise opportunities can be inculcated. One exercise that can be used to improve our capability to see the positive side of events is brain rewiring, this is where we purposefully record 5 things that we are grateful for on a daily basis. Students who took my courses were required to perform brain rewiring daily and by the end of the semester, most of them reported increased capability to see opportunities in their lives. This is going to be discussed in more detail in the next chapter, Emotional Intelligence.

2.3.2   Language and Thinking

It is interesting to note here that computer science has contributed a fair share of our current understanding of how thinking takes place. In the course of attempting to build computers that imitate the human brain, engineers and computer scientists needed to understand how the brain works in the first place. A number of theories have been developed. The thinking theory that I shall describe here is my favourite and it is congruent with the concept of influencing myelin growth. The theory elucidates the role of language in the formation of thoughts and it has implications on how we learn and communicate. Thinking happens in three stages:

1. Naming
2. Classifying
3. Wiring (Neuron Myelination)

In order for the brain to store and process ideas, thoughts and concepts, the brain starts by assigning these concepts names first. We do that all the time, when we meet new people, even if we do not know their names, in our mind we assign them some sort of a name. If the person is tall, we may name him the tall man, if she is not from this country, we may name her the foreign lady and so on. This facilitates the storage part of the thinking. Now, if we are to have an interest in the person, object or the concept, our brain shall attach some classification to it. This classification could be anything like good, bad, hot, bitter, difficult, etc. This will result in electrochemical activity in the neurons (firing) which will result in creating a connection in the brain. The interesting thing about this theory is that it provides some control over the entire process of thinking through altering the naming and classifying stages. Let me illustrate this with an example that I use when I train executives. When I show a picture of a cockroach and ask the audience to name it, all of them will say it is a cockroach. So this is the naming stage. When I ask them what they think of it, most of them say things like dirty, disgusting, ugly, etc. This is the classifying stage. Now when they are asked of what should be done with the cockroach majority will recommend destroying it because they are 'wired' to do so.

A Cockroach. (Source: www.kaneexterminating.com)

After explaining the thinking theory, I show the picture of the cockroach again and ask the audience to name it without using any negative terms. This implies not using the term cockroach as it has negative connotations already. One suggestion is to use its scientific name 'Blattaria.' This is a fresh name with no classification attached. Now the audience is asked to describe the Blattaria using positive or neutral terms. Suggestions like 6-legged, brown, flyer, resourceful, flexible, protein, roommate and so on start to emerge. When I ask what we should do with it, almost everyone in the audience change their mind and they are not recommending destruction anymore. Clearly, using different naming and classifying terms creates new brain patterns that are wired through training and repetition, and this how is learning takes place.

This can have an immediate positive impact on our capacity to develop new positive mental habits that can help us Conceive solutions to the challenges we face. I have institutionalised this at the School of Engineering of Taylor's University where I used to serve as the dean. I have outlawed the P-word (problem) and replaced it wherever it appears in the curriculum with either "opportunity" or "challenge." To encourage my students to stop using the P-word, I created what I call the "opportunity note." Modelled after a dollar, I give each one of my students an opportunity note at the beginning of my courses, if they keep their promise of not using the P-word throughout the semester, I personally sign the opportunity note for them. I have also mailed opportunity notes around the world to my online students. Clearly, I do not have a way to watch my students 24/7 but creating a physical object (the Opportunity Note) makes the word come to life. As more students report great and enjoyable experiences, the culture takes roots. The opportunity note is shown below and I would like to recommend that you start removing the P-word from your vocabulary and influence your thinking process at the naming stage.

The Opportunity Note

Lao Tzu once said, "Watch your thoughts they become words. Watch your words, they become actions. Watch your actions, they become habits.

Watch your habits, they become character. Watch your character, it becomes your destiny."

It is worth noting that in some languages, like Arabic, there are two different words to represent the English word "problem", "Mushkilah" is the Arabic word for problem as a source of trouble or worry; and "Masaalah" is the translation of problem as a question raised for intellectual enquiry. Replacing the word problem with either "challenge" or "opportunity" is aimed at creating mental habits of seeing all "problems" as solutions in disguise that require unearthing and discovering.

Knocking a man out with one punch, also referred to as a "king hit", was recently causing an increasing number of casualties in Australia. In July 2012, a "king hit" caused the death of Thomas Kelly. Another victim, Daniel Christie, was in a critical condition after receiving a "king hit" on New Year's Eve of 2014. This has sparked a debate about the use of the term "king hit" which seems to glorify violence. Some activists together with the family of Daniel Christie are suggesting to change the term from a "king hit" to a "coward's punch." This renaming is meant to deprive this act of violence from the glory which its current name lends. Mike Gallacher, the NSW State Police Minister agreed with this proposal and encouraged the community and media to use the term "coward's punch" instead to help embarrass and shame offenders.

We live in a world that is made of our words. People can make our days by saying a nice word to us and similarly, hearing a harsh word can spoil our moods for a long time. The thinking model of naming-classifying-wiring is an invitation to cultivate a language that facilitates the realisation of opportunities as well as success, happiness, and potential fulfilment for everyone involved.

2.3.3   CDIO Process as a Mental Model

The process through which we go about doing our daily business has an impact on our mental models and how we think. While delivering corporate training to bankers and business leaders, I get to realise that the world of business can adopt some of the methods engineers use routinely. In order to leverage the power of the brain and overcome its shortcomings, engineers adopt the CDIO process which is a systematic and structured way to Conceive, Design, Implement and Operate products, services, and systems.

Although structured, the CDIO process is not aimed at limiting the creativity of engineers, on the contrary, it results in unleashing the engineers' creativity through assisting them in avoiding the loopholes and biases of the brain that can affect the quality of solutions they produce. The CDIO process is aimed at creating systematic thinking habits in the mind. Four chapters of this book are dedicated to discuss each stage of the CDIO process in detail.

2.4   Thinking and Reasoning

The Noble laureate Daniel Kahneman, in his masterpiece 'Thinking Fast and Slow', identified two systems of thinking, the fast one which is quick and intuitive and the slow one which meticulous and detailed. As mentioned earlier, the brain is a power-hungry organ which means that thinking is an "expensive" activity, so the quicker it can be done the better. Fast thinking is also necessary in threatening and dangerous situations where a split second decision, such as to run away from a snake or jump out of the way of a speeding car, is needed.

Now let us demonstrate the two systems of thinking. If you are asked a question like what is 2+2, you will quickly answer 4; this is the fast thinking system in action. Now if I ask you to mentally calculate the product of 17x34, you will be engaging the slow system of thinking, you will activate more areas of your brain, where you will calculate and store interim results, and use much more energy than the energy used to answer what is 2+2. Using the fast thinking system in intuitive situations to save time, effort and energy, or in life threatening situations to save lives, is all well and good. The real issue is when we fall into the trap of using the fast thinking system to tackle a challenge that requires engaging the slow thinking system. Take for example the following question adopted from the Cognitive Reflection Test prepared by Prof Shane Frederick. "A bat and a ball cost $1.10 in total. The bat costs a dollar more than the ball. How much does the ball cost?" If you have answered 10 cents, welcome to the club, you have joined the majority of 3000 university students who gave this wrong answer. If you take a moment to check your answer you will realise that the correct answer is 5 cents. What has happened is that your brain took the easy way and used the fast, inexpensive thinking system to resolve a challenge that required the use of the slow deliberate system. This example is not rare, and both professionals and ordinary people make countless mistakes every day just by trusting their gut feeling. The Cognitive Reflection Test is shown on the next page.

The Cognitive Reflection Test (CRT)

(Cognitive reflection and decision making by Shane Fredrick)

http://mitsloan.mit.edu/alumni/pdf/Winter08-InnovationAtWork.pdf

1. A bat and a ball cost $1.10 in total. The bat costs a dollar more than the ball. How much does the ball cost?

2. If it takes 5 machines 5 minutes to make 5 widgets, how long would it take 100 machines to make 100 widgets?

3. In a lake, there is a patch of lily pads. Every day, the patch doubles in size. If it takes 48 days for the patch to cover the entire lake, how long would it take for the patch to cover half of the lake?

2.5   Critical Thinking

It is clear by now that thinking is a complex process that happens both consciously and subconsciously, both fast and slow. As we have seen earlier, in its search for meaning, patterns and order, the brain can make mistakes. Left to its own mechanisms, human thinking can be exposed to biases, prejudgment, discrimination and prejudice. Critical thinking refers to the deep, intentional and structured thinking process that is aimed at analysing and conceptualising information, experiences, observations, and existing knowledge for the purpose of creating original and creative solutions for the challenges encountered. Thinking critically requires a positive open and fair mindset that is able to objectively examine the available information and is aware of the laid assumptions and the limitations brought about by them. Another feature of critical thinking is that it is systemic and holistic in the sense that while examining a proposed solution, it examines its impact and consequences on other parts of the system thus ensuring that a solution at one level of the system does not create challenges and difficulties somewhere else.

In this era of Grand Challenges, critical and system thinking as well as thinking holistically are prerequisites to addressing these challenges. Imagine the challenge of climate change, for example, before introducing any treatment in the system or proposing any solutions, we need to be able to comprehend the complexity and connectedness of different parts of the system and how an alteration in a part of the system may have an unexpected impact on another part of the system. To enable holistic thinking, it is usual to map out system network architecture and then study the impact of changes made at different parts of the system on the overall system. Another example of a complex system is the human body. The system network architecture of the human body contains numerous subsystems such as the nervous system, the circulation system, the digestive system, the skeletal system, etc. Now if we need to prescribe a certain medication to treat a headache, it is necessary to see the impact of this medication on other systems; and if any other system reacts adversely to the drug. Other examples of complex systems include the rainforest, the Internet, and the global financial system. Businesses and companies can also be represented through system networks with suppliers, customers, regulators and other direct and indirect stakeholders.

The Great Leap Forward!

In order to rapidly transform the Chinese society from an agrarian to a modern one, the Chinese government made a series of planning decision between 1958 and 1961, these came to be known as the Great Leap Forward. One of the planning decisions was the hygiene campaign against the "Four Pests" initiated in 1958. Chairman Mao conceived the idea of getting rid of mosquitos, flies, rats and sparrows. The sparrows made it to the list because they were responsible of eating grain seeds depriving the people from them. The design of the campaign against sparrows included destroying their nests and nestlings but most importantly it was based on preventing the sparrows from landing, forcing them to fly until died of exhaustion. This was implemented and operated by millions of people banging pots and pans to scare the birds. The operation was so successful that it drove sparrows and other birds to the edge of extinction in China. However, the Chinese government finally realised that the birds played another ecological role by eating a large amount of grain-eating insects, keeping their numbers under control. The net effect of the campaign reduced the yield of crops, forcing the Chinese government to abandon the campaign and import sparrows from overseas to re-establish the natural balance. The famine still resulted in the death of 30 million people. This is an example of failure to think holistically and in terms of systems, where introducing a change somewhere in a complex system often will have other effects in other parts of the system.

2.6   Knowledge and Memory

Research has shown that we learn better if we develop accurate mental models of how our brains work. We have devoted this chapter to describing thinking in its different forms, and the importance of having a positive growth mindset to achieve happiness, success, and overall well-being.

For the sake of completeness, it is useful for us to explore how memory works. There are two types of popular mental models for memory, the short term memory, and long term memory. Short-term memory, or working memory, is where we temporarily store data. Long-term memory is where we eventually store events that happened to us and the learning that we have achieved.

One useful way to visualise memory is to think of a computer. The computer has a large long term memory, the hard disk; and a much smaller working memory, the RAM (Random Access Memory). When we save a document, it is stored within the hard disk (long term memory). If we choose to edit it, the document is recalled into the RAM and a word processor can make the changes required. Once done, the document can be committed back to the hard disk (long term memory). Now if there is a power outage while you are processing the document, you will lose all the unsaved changes but you can rest assured that the original version is still available in the hard disk. You may ask why don't we do the edits in the hard disk itself, the answer is simple, while the hard disk is very stable, the RAM works way faster allowing very quick changes.

If you are asked to remember these words in sequence: smart, orange, metal, lunch, dust, school, book, you will probably be able to hold them in memory. As a matter of fact research has shown that our short-term memory can hold up to 7 items at a given time. That is why the phone numbers have seven digits in them, but if you are distracted, for example by trying to answer what is 14x27, the short-term memory will fall apart. Long-term memory can hold much more than that, there is practically no upper limit to how much long-term memory can accommodate. However committing knowledge to long-term memory requires effort and intentional mental work, and is often associated with creating mental models that make sense of the information to be retained.

Our brains find it very difficult to work with random, meaningless or unrelated information that does not fit together into a framework that makes sense. That is why we like congruent stories, and why, when we want to get children to remember the alphabet, we do it as a song, A B C D E F G.... The song helps transform a random string of data into a rhythmic meaningful song. The phone number of my childhood home was 7610042 (do not try to call this number now, someone else may answer you!), in order for me to help my friends keep it in their short-term memory and help them store it in the long-term memory, I chunked the numbers in the following manner 76-100-42. "76" represents the length of Halley's Comet cycle, "100" the maximum grade can be achieved in an exam, and "42" is the product of the first two digits, 7 and 6. This sequence helped create two things, firstly it reduced the number of items to deal with from 7 digits to three chunks making it easier to handle by the short-term memory; secondly, it created a story, or a mental model that makes it easier for the phone number to be stored in the long-term memory. Chunking and creating patterns out of seemingly unrelated or random strings of data are some of the techniques that people with super memories use to memorise large amounts of data.

Joshua Foer is a journalist who grew interest in the memory championships that are organised around the world where participants exhibit herculean memory feats by memorising card decks and long unrelated string of numbers. To understand things better, Joshua decided to compete in one of these competitions and spent a year preparing for that. He narrates the story of his participation and winning of the memory championship in his book 'Moonwalking with Einstein: The Art and Science of Remembering Everything'. In the book, he describes many interesting memory techniques including a technique called the Memory Palace. This technique makes use of an existing familiar memory framework and superimposes an unrelated list on it utilising a story. The familiar framework is normally a home or a place that the individual knows very well. To remember things you will need to mentally "place" these things at certain corners of the memory palace. Let me demonstrate this with an example.

Let us say that you want to memorise the following shopping list:

- Milk
- Soap
- Cereal

- 6 Cans of 7-Up
- Screwdriver
- Cat Food
- Fish

Close your eyes and in your mind go to your childhood home or any other home that you are very familiar with. Imagine yourself outside your home and imagine that there is a swimming pool that is filled with **Milk**. Clearly visualise this image in your mind. The image may sound strange, but that is even better. Now picture your favourite movie star bathing in the milk swimming pool using a bar of **Soap**. I shall leave it to your creativity to visualise this. While the movie star is bathing in the milk pool with the soap, imagine a truck full of **Cereal** dumping its load into the swimming pool. Leave this behind and enter into your home. As you open the door, **6 Cans of 7-Up** stacked behind it will fall around forming the figure "6". Look at this with amazement! Walk into your kitchen and put your hand into your pocket and get a **Screwdriver** and use this screwdriver to force open a can of **Cat Food**. As the cat food spills all over the place, a **Fish** will jump out of the fridge, comes to life and eat the cat food. If you follow the instruction you will realise that it is difficult to forget the shopping list this way. I have tried this exercise with my students and they were able to remember long strings of unrelated items. We did the exercise at the beginning of the lecture and I checked them again at the end of the lecture and they recalled the list very well. Obviously, when items are stringed into a story that is told in a familiar setup (the memory palace) long-term memory can be invoked relatively easily. This technique is very powerful with people reported to be able to memorise an entire dictionary using it.

1. The human brain has three main regions:
   Old brain   - responsible for survival instincts.
   Middle brain - responsible for processing emotions.
   New brain   - responsible for high level thinking and language.

2. We live in a world that is made of our own words. We can influence our mindset through the creative description of the situations we are encountering. The more we describe things positively to ourselves, the more positive the outlook of our thinking becomes.

3. Thinking critically requires a positive open and fair mindset that is able to objectively examine the available information and is aware of the laid assumptions and the limitations brought about by them.

Practical Takeaway
Stop using the p-word (problem) and replace it with either "opportunity" or "challenge" in your daily life. This way, you will be able to influence your mind to assume a positive outlook, thus increasing your creativity and innovativeness.

Chapter 2
Brainology: The Biology, Psychology & Technology of the Brain

Chapter 3
Emotional Intelligence

"If your emotional abilities aren't in hand, if you don't have self-awareness, if you are not able to manage your distressing emotions, if you can't have empathy and have effective relationships, then no matter how smart you are, you are not going to get very far."

Daniel Goleman

"We cannot tell what may happen to us in the strange medley of life. But we can decide what happens in us, how we can take it, what we do with it, and that is what really counts in the end."

Joseph Fort Newton

Holistic education and human development are integrated processes that aim not only at developing skills necessary for employment, but also at ensuring the emotional well-being of individuals. This often is referred to as training the heads, hearts, and hands. The good news is that the only part of the body that we need to train is the brain! Knowledge, emotions and even manual skills are all cultivated, developed and stored in the brain. The integrated holistic development of knowledge, skills, and emotional competencies will result in individuals who are not only ready for employment but also ready for life's challenges through being resilient, purpose-driven individuals who can fulfil their full human potential and help others achieve the same. Purpose-driven individuals have a positive growth mindset that gives them a belief that they can have a positive impact on life. This motivation is a necessary prerequisite for performing meaningful and successful CDIO processes.

This chapter draws on the writings of Daniel Goleman, especially 'Emotional Intelligence' and 'Working with Emotional Intelligence'. It is, however, not intended as a replacement for reading Goleman's excellent books, in fact, I highly recommend them. I taught a course called "Success: Achieving Success with Emotional Intelligence" to my engineering students in Malaysia in 2013. The course was also offered as a free Massive Open Online Course (MOOC) which more than a thousand students from all over the world took online. Participants from all over the world found the tools I introduced useful and to complement the theoretical framework described by Daniel Goleman.

I measured the emotional intelligence of the students who attended my course and compared it to that of a control group that did not attend my course. The measurement happened twice, at the beginning of the semester and at the end of it, where students need to complete a detailed questionnaire. The results were very encouraging with the students who completed the 18-week course achieving growth in all aspects of the emotional intelligence compared to the control group. In this chapter I will attempt to introduce emotional intelligence from a practical point of view and provide simple exercises that can help grow different aspects of emotional intelligence.

3.1   What is Intelligence?

Intelligence is a widely used word but we seldom stop and ask ourselves what it means. It has been defined in a variety of ways that are related to cognitive capacities in humans. Intelligence has been observed in other living creatures as well. Artificial intelligence is the capacity that is programmed into machines (computers) enabling them to respond to new situations and learn. Intelligence can refer to the mental ability to think, learn, recognise patterns, logically predict outcomes, and respond to a variety of stimuli. Steven Pinker, the author of 'How the Mind Works,' defines intelligence as the ability to attain goals in the face of obstacles by means of decisions based on rational (truth-obeying) rules. Intelligence Quotient (IQ) is the measure of the ability to comprehend logical, geometrical, and mathematical challenges. While IQ is a useful indication of future success, the challenges of the 21st century increasingly need other kinds of intelligence that IQ does not measure.

As explained in the previous chapter, the brain has three main regions, the old brain, which is dedicated to survival instincts; the middle brain, where emotions are processed; and the new brain, where rational high level thinking happens. We perceive the world around us through the variety of sensory signals that are relayed to our brains. Signals flow in and out of the brain through its lower part, the old and middle brain; which means that any signal that goes through the brain will be emotionally "flavoured" before reaching the new brain for rational processing. This is the reason we sometimes emotionally overreact to events and stimuli. Being aware of this emotional overdrive and being able to manage the impact of emotions on ourselves as well as others around us is called Emotional Intelligence.

Surviving and thriving in the 21st century, which is a key century that is riddled with grand challenges for the humanity as a whole, requires holistic development of individuals in both the cognitive and emotional domains. This is a prerequisite for nations' competitiveness, prosperity and security. Despite the wide realisation that national education systems ought to be geared towards the provision of holistic education, there is a gap between this realisation and what happens in reality at schools and universities around the world. In his book, the 'Global Achievement Gap', Tony Wagner identified 7 survival skills for the 21st century that are not systematically developed in schools today. These skills are:

1. Critical thinking
2. Collaboration across networks
3. Agility and adaptability
4. Initiative and entrepreneurialism
5. Effective oral and written communication
6. Accessing and analysing information
7. Curiosity and imagination

It is clear that in order to maintain nations' competitiveness and achieve holistic development of individuals and societies, it is paramount to adopt an integrated approach to inculcating both IQ and Emotional Intelligence, leading towards the development of human capital equipped with all the necessary survival skills.

3.2   Emotional Intelligence

Deep inside our hearts (or should I say brains!), we are all ultimately chasing happiness. We may be seeking success in the form of money or power, we may be seeking status or fulfilment, but if we dig deeper, often happiness is what we find as our fundamental motive. Now happiness is not easily define either, as it means different things to different people. The definition I like is that happiness is the quality of one's relationships. If you think about it, after achieving the basics in life, such as having health, food, shelter, what really matters is how good are our relationships with our loved ones, students, teachers, co-workers, drivers on the road, and almost everyone else. Collaboration and teamwork, which are key success skills in today's work, school, and life in general, are also functions of the quality of relationships with others. Emotional intelligence is an essential ingredient for nurturing healthy and productive relationships. It is the skill and ability to understand, recognise, predict and appropriately respond to emotions in one's self and in others, as well as in groups and teams. Daniel Goleman presented a neat and useful framework to describe the four aspects of emotional intelligence, namely self-awareness, self-management, social awareness and relationship management. The framework, together with a description of its main aspects, is discussed below.

	Self	Social
Awareness	**Self-Awareness** Emotional Awareness Accurate Self-Assessment Self Confidence	**Social Awareness** Empathy Organisational Awareness Service Orientation
Regulation	**Self-Management** Emotional Self-Control Transparency Adaptability Achievement Orientation Initiative Optimism	**Relationship Management** Developing Others Inspirational Leadership Influence Change Catalyst Conflict Management Teamwork & Collaboration

Emotional Intelligence Framework. (Source: Daniel Goleman)

3.2.1   Self-Awareness

The cornerstone of emotional intelligence is self-awareness. Emotionally intelligent individuals are aware of their internal state, capabilities and limitations and they are able to use language to describe how they feel in clear and precise words. These individuals can manage themselves better and can develop better relationships. According to Daniel Goleman, self-awareness encompasses emotional awareness, accurate self-assessment and self-confidence.

3.2.1.1 Emotional Awareness

Emotions are often what sets us in motion, and the basis of emotional intelligence is to be aware of our own emotions and to be able to accurately describe and express them. This is easier said than done though! One of the reasons for this is that the middle brain where emotions are processed is incapable of processing language. Try to ask people how they feel. I will bet you the answer will be words like "fine", "ok" or something along these lines. Being aware of our emotional state is the first step in cultivating emotional intelligence, and developing the language to describe our emotional state is a prerequisite for achieving the emotional awareness.

To help the students who registered for my online emotional intelligence course develop the language to describe their emotions, I requested that they report their emotional and relational states daily using specific adjectives in 6 different domains, namely Mental, Emotional, Relational, Spiritual, Vocational, and Physical, domains. I learned this framework from Jim Warner who coaches top executives and CEOs around the world and he starts his forum sessions with a "check in" where everyone reports her/his M.E.R.S.V.P. state. This is described in more detail in the table on the next page.

The process of reporting one's emotions daily was difficult initially, where the students found it almost agonising as they forced themselves to describe their emotions. As time went by, the process got easier and more enjoyable. Throughout the process, I assured the students that "all emotions are OK." The purpose of the exercise was to create awareness of our feelings, and not to deny them or suppress them.

Sample adjectives that can be used to describe how we feel.

Domain	Definition	Sample adjectives
Mental	Mind, intellect	thinking, sharp, focused, curious, open, blocked, challenged, questioning, confused, learning, growing, wondering
Emotional	Affective state of consciousness	happy, sad, fearful, disgusted, guilty, confused, aware, excited, satisfied, loved
Relational	Quality/state of relationships	connected, grounded, networked, blessed, separated, reaching out, supported, lost, misunderstood, betrayed
Spiritual	Bigger cause/meaning	blessed, grounded, assured, doubtful, searching, enlightened, driven, betrayed, disappointed, fulfilled
Vocational	Job, career or study	progressing, stuck, challenged, stretched, supported, driven, focused, confused, realistic
Physical	Body and health	healthy, strong, in pain, flourishing, recovering, refreshed, healing, renewing

3.2.1.2 Accurate Self-Assessment

In order for us to progress in life, it is essential to be able to know our strengths, weaknesses, opportunities and threats. This is often called SWOT analysis and performing it can have a profound impact on our accurate self-assessment and self-awareness. Strengths are positive internal qualities and capabilities an individual (or an organisation) has. These can be physical traits, skills or qualities of character that can help in achieving goals. Weaknesses, on the other hand, are internal shortcomings that need to be addressed further in order to ensure the achievement of objectives. Opportunities and threats represent external events and circumstances that can potentially be useful or harmful, respectively.

As we perform the SWOT analysis, we can have a more accurate self-assessment, allowing us to capitalise on our strengths, use our weakness as opportunities for growth and be ready to grab opportunities, and face challenges.

Self	External
Strengths	**Opportunities**
Weaknesses **(Areas for Growth)**	**Threats** **(Challenges)**

Benefit / *Harm* (row labels)

SWOT Analysis

3.2.1.3 Self-Confidence

Armed with sharp awareness of one's current state along the 6 M.E.R.S.V.P domains and accurate self-assessment, a clear picture of an individual's potential as well as limitations emerges. This results in confidence of one's (and even organisation's) abilities to tackle challenges as they come along.

3.2.2 Self-Management

Managing self is a very important component of emotional intelligence. Individuals who are self-aware learn that all emotions are ok. It is ok to be angry, sad or happy. However, not all actions are ok. It is not acceptable, for example, to insult others or physically harm them while we are angry or disappointed.

We will always be exposed to external events and stimuli as we go through the ups and downs of life. Angry motorists will shout at us, co-workers will misunderstand us and people will cut the queue ahead of us. We need to accept that we will never be able to control external events that happen to us. What we have control over, however, is how to respond to these events and this is called self-management. An angry motorist can push our buttons, and this would surely make us angry and frustrated, but we still can choose the way to respond to this stressful situation. We can smile and apologise to diffuse the situation or shout back and escalate the situation to a fight. According to Daniel Goleman, self-management entails emotional self-control, transparency, adaptability, achievement orientation, initiative and optimism.

To survive in the environment in which our ancestors lived, our brains evolved to quickly respond to negative stimuli and threats. If a caveman is walking in the jungle, it makes sense for him to run if he feels danger approaching rather than stopping and analysing the consequences. This fast action that served us well in the past has become a liability that we need to mitigate in today's world. There is a need to train our brains to be able to identify and quickly respond to opportunities, not only threats. This fits clearly in the realm of self-management. Below are the techniques and activities that I introduced during my Massive Open Online Course to help cultivate different aspects of Self-Management components. Students found these techniques and activities to be beneficial and even life changing.

3.2.2.1 Brain Rewiring

As we repeatedly mentioned, the human brain seems to be hard wired to respond to negative stimuli. This trait was very important to ensure survival in the dangerous environment that our ancestors used to inhabit. However, this inherited asset may become a liability when it comes to situations where positive thinking and the ability to respond to opportunities are paramount. This is especially true when the task involves Conceiving, Designing, Implementing and Operating value-added innovative systems, products and processes. The good news is that the brain can be "rewired" to respond to positive stimuli.

One technique proposed by Ben Shahar to achieve this is to keep a gratitude journal reporting things that an individual is grateful for daily. I have institutionalised this exercise in my Emotional Intelligence course where all my students were required to report 5 things that they were grateful for on a daily basis for the period of 18 weeks. Just like reporting the inner state daily, the Brain Rewiring exercise started awkwardly and difficult, as it required different brain muscles to be trained. With time, the exercise became more fluent, easy and enjoyable. At the end of the course, students reported a more positive perspective towards life and appreciated themselves and those around them more. This positive mental attitude is extremely powerful when addressing challenges and it can cultivate the creation of innovative alternatives in both professional and personal life as well as fostering emotional self-control and optimism. Repeating this daily exercise for the period of the course was aimed at creating positive thinking attributes and literally building myelin along the mental circuits that enable us to see the goodness in situations and people.

Towards the end of the course, Susan, one of my online students, reported the following as one of the things that she was grateful for on January 11, 2014 "a really difficult conversation which brain rewiring enabled me to turn into a relationship building opportunity rather than a confrontation! Yeah!"

As mentioned earlier, brain activities governing behaviour, habits, learning and thought expressions occur as electric signals generated and transmitted along neurons in the brain. Neurons in the brain are widely connected and that makes the electric signals leak as they move along this complex network of wires. Performing tasks repeatedly and intentionally, such as brain rewiring, will get the brain cells firing signals in a certain repetitive fashion. Continuous firing stimulates the formation of the electrical insulating myelin sheath that encloses the path of the electrical signal. The more the neurons fire along the same path, the thicker the myelin layer and the better the insulation. In time this will improve the strength and the speed of the electrical signals and this is how we get better at things after practicing them.

3.2.2.2 Personal Vision and Mission

Having written mission and vision statements are very important. These days most organisations have these statements crafted and displayed on their walls. If done right, the vision and mission can provide a powerful tool to outline and communicate the purpose and core values of an organisation, or an individual. A mission statement provides the description of the purpose or "why" the individual (or organisation) exists. Individuals can have more than one mission, for example, a professional and personal mission, these will describe the professional purpose and the family purpose of the person respectively. Missions can be outward or inward looking. Usually an inward looking mission focuses on personal benefits and it is difficult to inspire others to join and support; an example of an inward mission "to be a millionaire by 2020". Outward looking missions usually have impacts beyond the individual and they can resonate with others who may support and help. An example of an outward looking professional mission for a teacher is "to help all my students realise and unleash their full potential".

A vision is a statement of what an individual (or organisation) wants the world (or themselves) to be when the mission is achieved. For example when the individual with the mission to be a millionaire achieves his (her) mission, he (she) can have "total financial freedom" and this could be his (her) vision. For the teacher who wants to unleash the students' potential, a vision may be "a world with no illiteracy" or "a highly competitive nation."

3.2.2.3 S.M.A.R.T Goals

In order to for a person or an organisation to achieve their mission and vision, planning and goal setting is a key activity. While mission and vision may take a lifetime to achieve, goals need to be set and achieved in a more foreseeable timescale.   In order to ensure the achievement of goals that accumulate towards mission achievement, goals need to be Specific, Measurable, Attainable, Relevant, and Time-bound. Below is a quick review of each aspect of the SMART goals:

Specific Goals

A specific goal is clear and not ambiguous. It needs to indicate who is involved in achieving it, what needs to be accomplished, where will it take place, when will it happen and how will it be achieved. For examples, to "Get in shape" is rather a general goal while "join a gym and work out 3 times a week for a minimum of 1 hour at a time" is a specific goal.

Measurable Goals

Here we need to specify the metrics for measuring the goal so that we can clearly know when the goal is achieved. For example, "increase sales" is not measurable while "increase sales by 10% compared to last year" is measurable.

Attainable Goals

For goals to be attainable there should be no ethical or legal barrier against them. When setting goals, one also needs to acknowledge the environmental and physical limitations. For goals to motivate, they need to represent a stretch on what is perceived as being possible while simultaneously ensuring that the skills, capabilities and attitudes that make the goals attainable are within the reach of those who are working on them. For example, if I do not know how to swim now, setting a goal of winning an Olympic medal in 6 months is not realistic and may demotivate me as I go about my weekly training. However, the goal of qualifying for a local swimming competition within a year, is a very motivating goal; even though it stretches me, if I train long enough, I have a very good shot at achieving it.

Relevant Goals

The goals we set ought to be relevant and congruent with the overall mission and objectives of the team or the organisation. A football coach, for example, may set the goal of "making 2 egg sandwiches by 9 am". The goal is specific, measurable, attainable and time-bound but it is hardly relevant if the overall objective is to increase the tactical skills of the team. Relevant goals are worthwhile, they add real value and capitalise on the strengths and capabilities of the team members.

Time-Bound Goals

The goal need to have a clear timeline indicating when they will be achieved.

While setting SMART goals is a very important activity, it is essential here to remain flexible, agile and open to review the goals should an interesting unforeseen opportunity arises. If we work in the mining business and while digging for silver, we find gold, we should have the mental agility to change our goals to capture the new "golden" opportunity.

When I worked on my PhD research, my project focused on exploring the relationship between fluid mechanics and the formation of gallstones in the gallbladder. The work involved building models of the cystic ducts of patients who underwent surgical removal of the gallbladder and studying the flow inside them. While observing the flow structures I noticed that there was some fluid mixing occurring within the cystic duct that connects the gallbladder to the rest of the biliary system. While, strictly speaking, my SMART goal was to explore the relationship between gallstones and the mechanics of the flow, I opted to pursue the mixing opportunity as a viable industrial option. This ended up being a significant part of my doctorate work and even influenced the title of my thesis which I changed to be "Cystic Duct to Static Mixer: A Serendipitous Journey." Let me iterate again, setting

SMART goals is very important for successful planning. However, maintaining an open mind and preserving intellectual agility allows us to adjust the SMART goals to yield the best outcome should there be a change that was not foreseen when the goal was set.

3.2.3   Social Awareness

Just like self-awareness, social awareness refers to presence of the mind and other character traits that enable individuals to be aware of their surroundings and how they are impacted by and impact other people's feelings and emotions. To nurture social awareness, Daniel Goleman proposed the development of empathy, organisational awareness and service orientation.

3.2.3.1   Empathy

Empathy is the ability to put ourselves in the position of others and feel how they feel. It does not necessarily mean to agree with others, just to feel what they feel. This is extremely necessary for all social animals and it has its roots in the biology of the brain. Italian researchers discovered certain neurons in the brain that are activated not only when we go through an experience but also when we see someone going through a similar experience. That is why, for example, if we see someone accidentally hitting her finger with a hammer, we almost feel the pain in our own fingers. These neurons are now dubbed "mirror neurons."

Empathy is a very powerful tool and those who cultivate it can be good team players. Empathy is very necessary for professional success as well. Engineers and designers, for example, need to develop empathy with the end users of their products so that they can develop human-centric designs.

Studies of the history of malpractice litigation against doctors revealed that empathetic doctors who exhibited care and listened well to their patients were at a much lower risk of being sued for malpractice by their patients when something goes wrong, compared to doctors whom were perceived as less empathetic. Empathy is also a key skill for those who work in sales, marketing and customer service.

One practical way to develop empathy in product design is to use the product through a process that is called "body storming." Imagine a group of male engineers who are designing a car that is being pitched to female customers. The body storming entails the male engineers put on high heels, carry handbags and try to get into the car and drive it. The knowledge acquired through body storming is very valuable and often ends up profoundly influencing the design process. After I turned forty, I started to have challenges in reading fine print. This includes the numbers and letters on my TV remote control as well as some name cards. Although I think that, at times, the fine print is made fine on purpose to fool consumers, but in the case of my TV remote control and name cards, the designers are either young or used enlarged models of their design on their computer screens. In both cases, putting themselves in the shoes of different age groups of their end users would have significantly improved the design.

When my colleagues who are new to teaching ask me how to know if the time allocated for their final exam paper is sufficient, my advice is to choose a quiet corner and answer the exam paper as if they are sitting for it themselves while timing themselves. The time allocated for the students should be 1.5 times the time the lecturer needs to complete the paper.

To develop empathetic skills, students in my Emotional Intelligence course were grouped in pairs. Each one of the two students was asked to describe his (her) feelings while the partner listens empathetically and in a non-judgmental way. The partner then reciprocates and listens back. After that exchange, each partner narrates the feelings of his (her) partner back to him (her).

3.2.3.2 Organisational Awareness

Organisational awareness refers to the ability to sense the dynamics of a group of people or team. This includes at which stage the team is in its growth cycle as well as the centres of power such as who calls the shots and who may be playing a negative role in the team. This awareness allows the individual to gauge the level of emotions in a group as well how can this be positively influenced.

3.2.3.3 Service Orientation

Service leadership is being recognised as a sustainable leadership style. Leadership is influence and servant leaders can command the respect of others and have a positive influence over them, enabling their teams to achieve their goals. To develop service orientation, it is important to get involved in activities that go beyond the direct benefit of an individual and have positive impact on others. Through serving others, we can develop a sense of self-worth and usefulness while building our capacity to understand the plight of those around us. Gradually, Service Learning is being acknowledged as a viable development technique for building emotional intelligence traits in students.

John-Son Oei completed a degree in communication. After graduating, he was searching for his life purpose and inner calling. "I chose to study communication because I thought it was a generic degree that will help me find what I should be doing later in life. However, after graduating I was even more confused." John-Son told my students while delivering a guest lecture as part of one of my MOOCs (Massive Open Online Courses). While enjoying life and making money modelling, John- Son felt the urge to do something more meaningful.

Travelling one day with his friends, he visited a village of one of the Malaysian indigenous people groups (called Orang Asli) and he was shocked to know that some of them did not have a proper home to stay in. He was particularly disturbed seeing a broken shack with a man living there. Something inside him made him feel that he should act! He posted a note on Facebook inviting his friends to join him to build the man a house. He was surprised by how many people were willing to help. They spent a weekend building the house and even gave it a paint job. Encouraged by what a group of untrained but enthusiastic people can do, John-Son went on and founded EPIC Homes, a social enterprise that builds homes for the Orang Asli.

EPIC Homes has a very interesting business model. It converts the home building exercise into a team building challenge that companies and organisations can perform over a three-day period. The company will pay for the materials needed to build the house and EPIC Homes will supervise the team from the company throughout the home building process. The home is then presented to the identified Orang Asli family. This service learning exercise is a very interesting and innovative initiative that EPIC Homes is pioneering. It takes place in partnership between the beneficiary family and the team which is building the home. "Poverty is a state of mind," John-Son says. "In order to empower the family that will receive the home, EPIC engages the family members in the design of their future home. They can choose one of three available designs. They are also given a choice of what colour the paint will be," he added. To ensure that the beneficiary family does not see the house as free hand-out, the family members are also required to participate in building their house and also to pay-it-forward and help build two other homes with their neighbours. This had a very interesting spill over, as members of the Orang Asli community participate in building their home, they develop very useful skills that helped some of them gain employment. EPIC Homes has built 23 homes thus far, a small dent in the 12,000 homes needed by the Orang Asli community according to John-Son.

John-Son Oei, Founder & CEO of EPIC Homes

I had the honour to be part of the Taylor's University team that built a home for Juri and Masni and their three children. The project was sponsored by Taylor's University, the senior management team of the university, supervised by the EPIC Homes team, worked very hard to build a nice and simple house in three days. Service orientation is an important and necessary trait to develop Social Awareness through feeling the needs of others and being able to play a positive role in fulfilling these needs.

Building a Home for Masni and Juri (Mushtak Al-Atabi)

3.2.4   Relationship Management

While self-awareness is the cornerstone of the emotional intelligence, relationship management is its ultimate objective. As mentioned earlier, happiness is measured by the quality of relationships that we cultivate with those around us. The pillars of the character of an individual who is capable of nurturing and managing great relationships with others include developing others, inspirational leadership, influence, being a change catalyst, conflict management and teamwork and collaboration. These traits are discussed below.

3.2.4.1 Developing Others

A distinct sign of emotionally intelligent individuals and leaders is a genuine interest in developing others. An exercise we did to inculcate the skill of developing others in the students who took my Emotional Intelligence course was to pair students together, and get them to interview each other to document each other's mission, vision, SWOT analysis and a SMART goal to be achieved by the end of the semester. Each student was encouraged to conduct the interview with care and respect and carefully write down the mission, vision, SWOT and the SMART goal of his (her) partner.

At the end of the exercise, each student would have to face his (her) partner and say their name followed by "your SMART goal is _____ and I know that you will achieve it." Students are also asked to send each other specific encouraging messages related to their SMART goals twice a week. This can be done verbally, using email, text messages or even with a card.

After completing the exercise, one of my students posted the following on the course website:

"Today I would like to share something that I should've shared a few weeks back. So during one of the lectures, we conducted an activity called mission partnership. I was actually paired up with my best friend. But eventually I was paired up with another person. A boy from Kazakhstan named '.......' At first I was dreading it and thought that I was going to fail the activity due to his lack of English communication skills and his 'blur'ness. But after starting the activity, my eyes were opened. I could see from his perspective, I was able to see what I could do. I was able to see the difficulties he was facing. And it broke my heart. Judging before

doing. And I would just like to thank Mr. Mushtak for that class and the activity and like to say sorry to '.......' This activity has opened my eyes. And I urge my fellow colleagues, don't judge and get into the person's shoes."

3.2.4.2 Inspirational Leadership

Leadership is a human quality that is both important and difficult to define at the same time. I would like to define leadership as the ability and willingness to exhibit initiative and take responsibility. Leadership also entails having a vision for a better future and readiness to serve others. It is important to remember that leadership is seldom about the position. Anyone who has a vision for doing things better, to serve others and improve their life and who takes initiative to bring along the positive change while taking responsibility for her (his) action can be described as a leader. I see this definition of leadership as liberating as it implies that anyone of us can practice leadership without waiting to be officially appointed.

For those who hold an official position of leadership, the expectations are clear. They need to serve, show the way and be ready to take responsibility and face the consequences of their actions and decisions. They need to be ready to take blame for failure and share credit for success.

In 2013 our school went through a major accreditation exercise. To successfully complete this exercise, academic staff need to participate in a very tedious documentation process. The time commitment to perform the documentation is not trivial and academics around the world dread it and view it as a necessary evil. We thought of a way to transform the staff experience and achieve genuine buy-in and we came up with a process that we called "Shoot the Boss!" It involves providing the staff with workshops and training sessions over 6 months to prepare them for the accreditation exercise. This included a mock accreditation exercise performed by experienced auditors. Staff were informed that the management have full confidence in them and while the success of the accreditation exercise depends on them, the management is taking full responsibility for any failure. The process concluded at a paintball court nearby, where the entire staff force attended and each one was asked 3 questions to assess their level of preparation and readiness. I informed the members of staff that if anyone of them fails to correctly answer at least 2 out of the 3 questions, that person is required to shoot me with the paintball marker as this is a sign that I have failed to give them the proper preparation. As a dean, I wanted to show the staff that I am taking full responsibility for the accreditation exercise. Generally that staff did very well, and one lecturer was emotional when she missed the correct answers of some questions as she thought she answered correctly and did not want to shoot me!

After the "Shoot the Boss!" session we had a couple of paintball games and went for great lunch together. Two months later, we had the actual accreditation visit and I was really touched to receive the accreditation report. Not only it was positive with all our programmes being accredited, but with the accreditation committee recognising "highly motivated staff and students" as one of the strengths of the school. Now whenever alignment is needed, people say we need to "Shoot the Boss!" What I learnt from this session is that when the leader is willing to serve, communicates the vision clearly and shows full accountability, people are more than willing to support and pull together and even the most tedious task can become enjoyable.

Shoot the Boss!

3.2.4.3 Influence and Change

Emotionally intelligent individuals can positively influence those around them and be agents of constructive change in their environments. One such individual is Jack Sim, the founder of WTO (World Toilet Organisation). At the age of 40 and after staring 16 successful businesses, Jack Sim had what he called a mid-life crisis where he started to question his own purpose. Realising that millions of people around the world do not have access to proper toilets, Jack decided to do something about it. "I was an academic failure as I had no degree. I also had no status and no authority," Jack said. "Starting on shoe-string budget and the hope to change the world, I had no choice but to attempt to enact change through influencing others to act," he added. Jack calls his framework of change and influence O.P. (Other People). The idea is very simple, if you are clever enough and have a worthy cause, you can leverage unlimited resources to your cause; these include Other People's money, talent, authority, time, power and brand.

Realising that there were a number of Toilet Organisations around the world, Jack founded the World Toilet Organisation in Singapore in 2001. He managed to convince various organisations that it is a good idea to have their HQ in Singapore. From the start, Jack used humour to break the taboo associated with toilets. This started with him playing on the WTO acronym, which belongs to the World Trade Organisation and did not end with him getting photographed wrapped with toilet paper and carrying a plunger. Leveraging media, business leaders, politicians and even Hollywood and Bollywood superstars to the cause, Jack was able to help many communities around the world, especially in India and Africa. In 2010, the World Toilet Organisation established the SaniShop which managed to sell 5,000 toilets thus far. The business model of the SaniShop is based on enabling communities to build their own toilets and improve sanitation and health in the process. The United Nations recognised Jack Sim's effort and currently 19 November is celebrated as the World Toilet Day.

Jack continues to influence people (Other People, as he called them) to act. His latest O.P. leverage is the production of a full-length feature comedy film "Everybody's Business." The story of this Lee Thean Jeen movie revolves around 50 Singaporeans getting food poisoning because of toilet hygiene issues. The fictional Minister of Toilets, together with hygiene officers of the Ministry of Toilets, goes around trying to reach the bottom of the matter.

Once again, Jack manages to use humour to break the taboo and delivers the message.

When Jack spoke to my students his advice was very simple, if you know what your purpose is and believe in making a difference, then think of the abundance in the world around you. If you align Other People to your goals, the cause will always be the winner.

Jack Sim. Founder of World Toilet Organisation.

3.2.4.2 Managing Conflicts

As more and more human activities involve working with people from different backgrounds, cultures and nationalities, conflict may be inevitable and it can range from argument, disagreement, and emotional tension to fighting or war. The conflict often starts with a misunderstanding or disagreement, if parties agree to disagree, there will be no conflict. In the book 'Hostage at the Table' by George Kohlieser, the following sources for conflict are identified:

1. Differences in Goals
2. Differences in Interests
3. Differences in Values
4. Differences in Communication styles
5. Differences in Power and Status
6. Insecurity
7. Resistance to Change
8. Role Confusion
9. Search for Ego Identity
10. Personal Needs
11. Poor Communication

To effectively manage conflicts, a bond between the conflicting parties needs to be maintained at all times. Couples with different religion, racial or political orientations can keep the bond and manage any conflicts that they may have. The misunderstanding that causes disagreement and conflict can stem from misunderstanding of one's self and misunderstanding of others. If we do not know what we want or what others want, or if we do not know how we feel or how others feel, misunderstanding can happen and it can lead to conflict. By now you can imagine that self-awareness is the antidote to misunderstanding of self, and social awareness is the antidote for misunderstanding others.

Practically, you can use the approach below to resolve conflict as described in 'Resolving Conflict Creatively' by Linda Lantieri:

1. Calm down, tune into your feelings, and express them.
2. Show a willingness to work things out by talking over the issue rather than escalating it with more aggression.

3. Try to find equitable ways to resolve the dispute, working together to find a resolution that both sides can embrace

Success Stories

Students who took my Emotional Intelligence course have numerous winnings to report. Below are some of them.

Michael found speaking in public to be a challenge, possibly because he has had a stammering issue since childhood. This is an account of how, with the help of the activities in the course, he overcame his public speaking challenge.

"The other day I had to give a speech, initially I thought of backing off. But this time instead of focusing on (the) stammering I focused on the positives, like people are there to listen to the content and things like that. I just prepared the content well and just imagined myself speaking with confidence and articulating each word slowly. The moment I went and faced the audience, the initial fear was like eagles pounding on me, but when I spoke the first word, those fears vanished into the woods. Somehow the magic happened, (I had) a sense of the "I can do it" attitude and I spoke, even to my astonishment."

He went on to say that through this experience and the emotional intelligence MOOC, he developed a positive attitude towards public speaking.

"...It was the same me, the same stammering and the same public speaking... But what I have learnt in the course is that, when I change the negative feeling of stammering to positive one, magic happens. Thank you Prof Mushtak for touching my life."

Andrea has confessed to having a short temper and she mentioned that ever since she started following the course, she found herself looking for hidden opportunities in every incident that would make her angry and then smiling at the thoughts. Now and again when anger still appears, she approaches it differently. She narrates this incident of her and her daughter.

"...like tonight when I got MAD at my daughter for not doing what I had asked her (my daughter is 3 and apparently is on a mission to test

me, provoke me and try my patience ☺). I did scold her, but the minute after I managed to compose myself again and explain to her my feelings and reactions. I was open to communicate with her whereas in the past I would not have been willing to listen to what she had to say because I was feeling my anger too much! We hugged and kissed and agreed that we will always be friends and talk ☺ Granted, about 30 minutes later she embarked on her mission again and there it was, enter anger and madness. This time I didn't get to explain because she had fallen asleep, but will do it tomorrow. I was happy that I could talk to my husband about this, without still feeling angry and without snapping and yelling!"

Highlights

Emotional Intelligence is a very good indicator for success in life. Emotional Intelligence has 4 main domains that can be improved with intentional practice. These domains are:

1. Self-Awareness
2. Self-Management
3. Social Awareness
4. Relationship Management

Practical Takeaways

1. To develop Self-Awareness, on a daily basis practice reporting how you feel in the following domains:

    1. Mentally
    2. Emotionally
    3. Relationally
    4. Spiritually
    5. Vocationally
    6. Physically

2. To rewire your brain to respond to positive stimuli and realise opportunities, keep a Brain Rewiring journal where you report the 5 things that you are grateful for on a daily basis.

Chapter 3
Emotional Intelligence

Chapter 4
Conceive

Conceive /kənˈsiːv/ v.
Create (an embryo) by fertilising an egg
Form or devise (a plan or idea) in the mind
Form a mental representation of; imagine

Oxford Dictionary

"Imagination is more important than knowledge."

Albert Einstein

*"Whatever the mind of man can **conceive** and believe, it can achieve."*

W. Clement Stone

As mentioned earlier in the book, every human-made artefact somehow starts in the brain. This is true whether we are talking about a consumer product, a mobile phone app, a bridge, or an airplane. This is also true for other less tangible products such as songs, jokes, symphonies and novels. As a matter of fact, the number one thing that differentiates humans from other beings is our ability to imagine and create in our minds, things that are not present at the moment or even yet exist. The defining factor (and the limiting factor as well) in enabling us to build homes, roads, cars, and all the other wonderful things that we have today is not that we have, arms, hands, fingers, legs or eyes; but the fact that we are able to imagine and Conceive, in our minds, what is yet to exist. Even animals, such as the great apes, that exhibit intelligence and primitive use of tools, and some insects that can build complex structures fail to progress towards developing a civilisation, principally because they lack imagination, which is a main ingredient of Conceiving.

Illustrating this point, we can cite numerous examples of individuals who, despite lacking fully able bodies since birth, managed to defy all odds and create new realities for themselves and the people around them. Richie Parker, for example, was born in South Carolina on May 1983. His father Tracy Parker was "dumbfounded," he told reporters years later, when the doctors told him that his baby is born without arms. Nonetheless with the help of his family, Richie was able to do all the things a normal child would do, including riding a bicycle and he ended up working as a vehicle design engineer at Hendrick Motorsports, one of the most successful companies in the NASCAR business. If this is not impressive enough, we can take the example of Nick Vujicic who had no limbs when he was born in 1982 and ended up becoming an internationally acclaimed author and motivational speaker.

Nick Vujicic (Source: Wiki Commons)

The premise of this book is that we have the capacity to create and engineer our own frame of mind and mindsets, as we perceive the world and imagine the future, and this is the basis for human creativity and potential. It is therefore extremely important to cultivate a positive growth mindset that harbours the conviction that every challenge, no matter how grand, has a solution that is waiting to be discovered and delivered and that there is no limit to what humankind can achieve. The true limit exists only in our minds and it becomes a reality once we believe that something is possible or impossible. From an engineering perspective, Conceiving represents a very deliberate and intentional stage of the CDIO process that can be learnt, and improved through systematic practice and coaching. While Conceiving (also called ideation) is directly connected to thinking, it is not only thinking. Instead, it is the use of the power of the positive mindset as well as knowledge and thinking skills to generate ideas and concepts for products and processes that can add value. It is worth reiterating here that the quality of ideas Conceived is affected by the mindset as much as they are affected by the mastery of the thinking skills and techniques. Conceiving also requires deep understanding of the nature of the challenge to be addressed or the potential of the opportunity to be realised. As demonstrated by Professor Shane Fredrick in his bat and ball example mentioned earlier, it is rather easy to fall into the trap of the illusion of knowing the answer to a question or a challenge before really understanding it. Hence, the systematic way in which the Conceiving stage is structured here will protect us against falling for this illusion when following the rules.

4.1   Preparing to Conceive

Driven by the need to improve an existing situation or the hope of realisation of a foreseen opportunity, and armed with their technical knowledge and understanding of the existing paradigm, individuals can start using systematic techniques to generate ideas. Before discussing the various techniques available in the engineers' toolbox to generate ideas, it is useful here to set the rules for successful Conceiving:

1. Clearly define and outline the challenge statement (or the opportunity statement) that is being worked on. If you are providing a solution to a client, the client requirements and desires as well as limitations need to be seriously considered and documented. This should be done in writing

and it requires deep understanding of the current paradigm and its promises and limitations.

2. Gather sufficient information about the challenge or the opportunity that you are working on. This includes reviewing the literature, online resources, and patent databases as well as performing market studies and surveys. Understanding the legal requirements, such as government regulations, represent an essential part of this information.

3. Keep an open mind and do not converge on a solution prematurely. Do not start with a pre-conceived idea of how to solve the challenge or realise the opportunity. This goes hand in hand with the first point when we define the challenge or the opportunity. For example, if we want to make a hole in the wall for a screw to hang a picture, the challenge statement should not be "to drill a hole in the wall" but instead it should be "to fix a picture on the wall." This clever framing of the challenge can send us in a direction of making some sort of fastener that does not even require making a hole in the wall. Without this, it is almost certain that we will end up with a drill to make the hole.

4. Always trust the process.

The above four rules are antidotes to the four temptations that engineers are often faced with. Falling for these temptations can render the Conceiving stage valueless, compromising the success of the entire CDIO process. These temptations are:

1. My team and I know what the challenge (or the opportunity) is. There is no need to write down the challenge (or the opportunity) statement.

2. What is in the literature is not relevant to the challenge I am facing. Or, there are too many things to search.

3. I know what the solution should look like.

4. The process is too lengthy; I do not have to follow all of it step-by-step.

As stressed in the Emotional Intelligence chapter, an important trait that needs to be developed to ensure successful Conceiving is empathy. As ultimately products and solutions we develop are to be used by other people, being able to empathise with others is an extremely valuable skill.

4.2 Ideation: The Art of Idea Generation

There are numerous thinking and ideation techniques available. These techniques vary in their degree of structure and suitability for group or individual settings. This section outlines a few techniques that I personally use and teach to both my students and the corporate clients. It is worth mentioning here that the corporate clients I train are not always engineers; they include bankers and governmental agencies' staff. If you are interested in reading more about these and other ideation techniques, you can refer to the work of Edward De Bono, Tony Bozan and Darrell Mann.

4.2.1 Brainstorming

Brainstorming is one of the most widely used (and abused) methods of idea generation. The term "brainstorming" is becoming ubiquitous and almost everyone is under the impression that they know how to brainstorm. Closer examination shows that people get very little training on how to use this technique. The brainstorming technique was first described by Alex Osborn. It can be used to generate ideas when working in groups of ideally 6-12 individuals. There are three main stages for a successful brainstorming episode, pre-session, session and post-session.

The pre-session includes clearly defining the focus of the brainstorming session. This should be identified and written somewhere so that everyone participating in the brainstorming session can see it. Whether you are brainstorming to find ways to increase sales, raise funds or develop an effective product, the first step is to define that focus. In the pre-session, a brain networking exercise will be very helpful in resulting in a fruitful brainstorming session. The brain networking is a warm up exercise that precedes a brainstorming session and is done as follows. Starting with a word, an item or a picture provided by the leader, a member of the team needs to mention the first word that comes to mind when seeing the picture or hearing the word, the next member needs to mention the first word that comes to mind upon hearing the word said by the first team member. The third team member needs to mention the word that comes to mind upon hearing the word of the second member and so on, this can go on for few rounds. You will notice that in the beginning, the process is a bit awkward (especially when done for the first time), and it has many pauses, this is to be expected, so do not worry and continue until the process is smooth and words are coming spontaneously. This is an indication that everyone is now on the same wavelength.

The brainstorming session can now start. The team needs to assign a scribe who will be writing all the ideas contributed by the team members on a board or a flipchart. It is necessary to always remember the rules of brainstorming which are:

1. One conversation at a time. The team is encouraged to use a talking item (like a small ball); only the person who holds the item can talk.

2. Stay focused on the topic. Check the ideas being generated against the session focus to prevent the group from sliding away from that focus.

3. Emphasise quantity over quality. Brainstorming is aimed at generating as many ideas as possible and as quickly as possible, so all ideas are welcome.

4. Encourage wild ideas. Wild ideas may not provide the solution you are looking for, but they can provide a gateway for a breakthrough into an entirely new innovative solution.

5. Postpone criticism of other people's ideas. This will encourage more people to contribute more ideas and lead to a richer brainstorming session.

6. Build on the ideas of others. When you hear an idea, try to use it and improve on it rather than starting a new non-related train of ideas.

Now that many ideas have been generated, the post-session will focus on selecting the best ideas for further processing. This is done through organising the ideas and combining similar ones. Every member of the team is then given three votes to vote for her/his favourite ideas. The ideas that received the highest number of votes are then selected for further processing and improvement.

A successful brainstorming session should produce a single idea (or a few ideas) that addresses the session focus that the team is very satisfied with and ready to work on and implement.

Ideas Selection and Classification

4.2.2 Random Entry

Random entry is a powerful technique that can help generate real breakthroughs in a divergent variety of situations; however, it requires some training to achieve the best results quickly. The random entry technique can be summarised as follows:

1. Clearly define the focus of the ideation session.

2. Select a random object to be at the centre of the process.

3. Identify the different attributes of this object.

4. Superimpose the object's attributes on the focus and see if it makes sense or presents an interesting idea line for thoughts to follow.

To demonstrate this technique, imagine that you are working for a company that manufactures computer printers. Your company needs to develop a breakthrough product that differentiates it from competitors and your boss asked you to come up with the product idea. Using random entry, you define you focus as "A new printing product" and the random object to be a flower. The traits of the flower include: colour, smell, pollination, thorn, garden, spring, etc... Superimposing the traits on the focus will result in things like:

1. Colour based/related printing product

2. Smell based/related printing product

3. Pollination based/related printing product

4. Thorn based/related printing product

5. Garden based/related printing product

6. Spring based/related printing product

Clearly, the colour printing has already been invented, but the smell based printing product is an interesting line of thinking to be pursued and developed further. So this may lead to a printer that can print pictures that have smell and developed further into a product which enables you to capture the smell of a place when you capture a picture and reproduce the smell when the picture is printed. Imagine scent cartridges next to the ink cartridges of your printer!

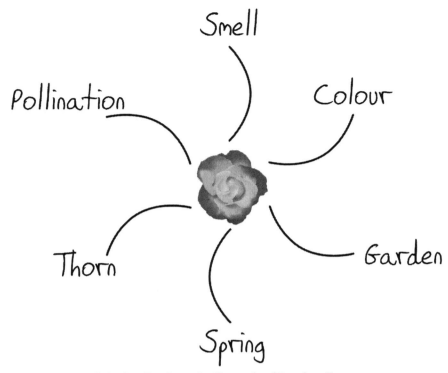

Smell

Pollination

Colour

Thorn

Garden

Spring

Printing Product: An Example of Random Entry

Random entry technique can be used individually (by one person) and in teams. Clearly, it is highly unstructured and the results are highly dependent on the object used to start the process as well as the selected attributes of this object. That is why this technique is recommended when we are running out of ideas to address the challenge and when we are mentally stuck. In this case arbitrariness, randomness and uncertainty can be rather liberating. If the first selected object did not yield a suitable idea, the process can be repeated either with a new object or with one of the attributes of the current object as the starting point.

Random Entry Case Study

Our economy is still, by and large, dependent on fossil fuel for energy generation. This results in huge amounts of carbon dioxide being released into the atmosphere, contributing to climate change. One way to mitigate this is carbon sequestration, where the emitted carbon dioxide is captured and pumped underground into bedrock deep underground. One of the Grand

Challenges for Engineering is to develop effective techniques for carbon sequestration. Using the random entry method, let's Conceive some new ideas to achieve carbon sequestration.

Carbon sequestration is generally done through removing carbon dioxide from the atmosphere and storing it in long-term reservoirs, normally underground. To Conceive some new ideas (ideate) to achieve this objective, let us first craft the statement that will drive the ideation process. As we are looking for new ways to store carbon dioxide we can try statements like, *"Carbon dioxide stored as _____"*

Then we need to pick a random object to drive the random entry process, let us pick a pebble and identify the different attributes of this pebble which may include:

1. Spherical - round

2. Colourful - colour

3. Hard - solid

4. Playable - joy

5. Collide with other pebbles-energy

6. etc.

Inserting each of the attributes into the statement above, we will have the following statement to try to make sense of:

*   Carbon dioxide stored or converted into something round- storing it into round shape (storage tank for example).

*   Carbon dioxide stored as colourful material.

*   Carbon dioxide stored as solid/hard object.

*   Carbon dioxide stored as game or a toy.

*   Carbon dioxide stored as energy

If none of the attributes gave a meaningful insight, we may use the attributes themselves as a source to expand the word list and inspire more ideas. Ultimately, we may use a different object altogether. In this case, converting carbon dioxide into a hard object seems like an interesting idea that can be developed further. What if we can make bricks that captures the carbon dioxide within them?

As mentioned earlier, the results of the random entry technique are clearly dependent on the random object selected to seed the process and the experience of the practitioners. But in this divergent nature of the technique lies its power. You may need to repeat the process using different objects before achieving satisfactory results.

4.2.3   Trimming

This process is suitable when working with an existing process, product or system with the intention of improving it. It can be performed by an individual or in a group and it has three primary steps:

1.   Break the process, product or system into its individual main components.
2.   Identify the functions of each components
3.   Mentally, select a seemingly important component (function) and trim it
4.   Ask yourself *"Now, how do I make it work without the removed component?!"*

This very powerful method can help us overcome some of our thinking biases. Many of the well-established products have earned themselves a permanent mental image: try to imagine a car, an airplane or a toaster and you will see how difficult it is to come up with a radically different version of them. When we are forced to remove (trim) an important component of the process, product or system we are improving, we put our minds in different frames that can enable breakthroughs.

An example of using the trimming can be demonstrated here. If you start with eyeglasses, you will notice that they have two major components, the lenses and the frame. Now if we trim the frame and try to make it work, inevitably we will end up with contact lenses.

Besides product design, trimming technique works for developing new ideas for business models and processes as well. Think of a bookstore, its business model has components such as publishers, books, shop, workers, payment system, etc… Now, let us try to trim the shop and think of a way to make the bookstore work. This will lead us to having something like Amazon.com, a bookstore without shops. Let us try something different, trim the publishers while keeping the business model operational. It is actually possible to have a new bookstore that works exclusively with the authors helping them to write and market their books directly to the reader. This bookstore may be serving a very niche and specialised market. Currently, services such as Lulu.com are performing a similar roles where authors can get their books printed on demand and sold without the need for publishers.

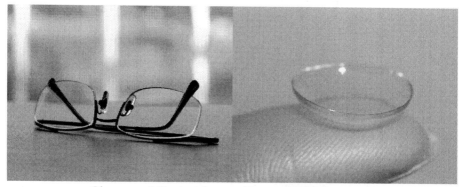

Glasses and Contact Lenses (Source: Wiki Commons)

Try to think of washing machines without water and a car without wheels and see how this can this lead to innovative new products.

4.2.4   Blue Ocean Strategy

W. Chan Kim and Renée Mauborgne of INSEAD developed the concept of the Blue Ocean Strategy, which refers to the development of new markets and new competitive advantages. In their book 'Blue Ocean Strategy', they discussed how businesses and organisations can make the competition irrelevant by creating new markets and moving away from the highly contested traditional value propositions, or what they termed "Red Ocean." They proposed a four-step technique to achieve that. The technique goes by the acronym ERIC (Eliminate, Reduce, Increase, Create). The technique looks at what the industry in which the organisation operates is providing and competing on and ask the following questions:

1. **Eliminate:** What is the organisation/industry providing now that can be entirely eliminated without affecting the value delivered?

2. **Reduce:** What is the organisation/industry providing now that can be reduced without affecting the value delivered?

3. **Increase:** What is the organisation/industry providing now that should be increased to enhance the value delivered?

4. **Create**: What is the organisation/industry not providing now that it should start creating to enhance the value delivered?

These questions entail studying what is available and adjusting what is being offered. The first and fourth questions, specifically, requires ground-breaking and original thinking to remove an offering that is seen essential or adding an offering that has never been considered before.

When Steve Jobs became CEO of Apple in 1997, he applied a similar process to reinvigorate Apple and bring it back to the main stream of the industry. First he entirely eliminated the printers and servers divisions of Apple. Second, to rationalise the wide range of products the company made, he drew a two by two matrix with the columns marked Personal and Professional, and the rows marked desktop and mobile. He challenged Apple engineers to decide the four products that will fit in the four quadrants of this matrix for the company to focus on. This led to the dramatic reduction of what is being offered and allowed the company to perform the third step of the Blue Ocean Strategy, to increase what really matters, which is the human-centric design and user friendliness of its products. The final step was to offer new products and services that were not traditionally seen as part of the industry's offering. In this, Steve Jobs was particularly successful as he introduced the company into the music business (iPod and iTunes), phone business (iPhone) and tablet business (iPad).

The technique can be extended to product design by rewording the four questions to be:

1. What features/parts of the studied product/process can be entirely eliminated?
2. What features/parts of the studied product/process can be reduced?
3. What features/parts of the studied product/process should be increased?
4. What additional features/parts should be created in the studied product/process?

The ERIC grid shown on the next page can be used when utilising the Blue Ocean Strategy to create improved products and processes. The example given here is for a budget airline.

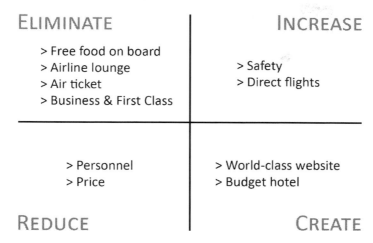

ERIC Grid for Budget Airline

The ERIC framework is aimed at strategically diverting resources from what is Eliminated and Reduced towards what is Increased and Created in order to produce new value and create an uncontested market. A highly useful tool to facilitate the creation and utilisation of ERIC framework is the "strategy canvas", which is a diagram with a horizontal axis that portrays the range of factors that the industry competes on, invests in and devotes resources to, while the vertical axis depicts the value offering levels that customers receive across all of these key competing factors. To develop the strategy canvas, the following steps can be used:

1. Identify the factors the industry is currently competing on.

2. Plot these factors on the horizontal axis of the Strategy Canvas.

3. For each "competing factor" indicate the industry benchmark "value offering level". This is done on the vertical axis with a scale that ranges from low to high. Connecting the points will give the existing value curve.

4. Identify what factors to Eliminate, Reduce and Increase in order to develop the new value curve.

5. Identify what new factors to add to the horizontal axis that no competitor is currently considering.

6. Connecting the value offering levels of points 4 and 5 will result in the Blue Ocean Strategy value curve.

To demonstrate the use of Strategy Canvas, let us study how a circus successfully deployed it to create new value curve. Circuses around the world are challenged by reduced revenues and competition of other forms of entertainment. A key component of circus spectacle, animal performance, is becoming a liability with high insurance and cost of care for the animals and with more people growing uncomfortable with the use of animals for entertainment. Cirque du Soleil is a Canadian circus founded in 1984 that responded to the challenges it was facing with Blue Ocean Strategy thinking. The circus wanted to reinvent the experience of its customers and provide them with a differentiated value by rebranding its shows and pricing them closer to the prices of a musical or theatre, which well beyond what customers are used to pay for a circus ticket. To achieve this, customers would expect a more refined and sophisticated entertainment experience.

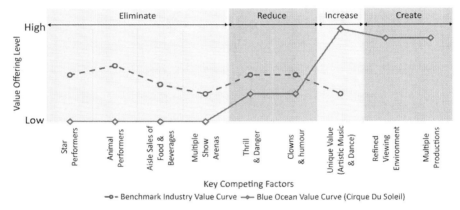

Blue Ocean Strategy Canvas for Cirque du Soleil

The circus management decide to Eliminate animal shows, the use of star performers, multiple show arenas as well as the sale of food and drinks in the aisles of the circus. To create a more sophisticated viewers experience, the circus Reduced thrill and danger as well humour in favour of more intellectual encounters. This resulted in an Increased unique and differentiated offering compared to competing circuses. To cement its position as a competitor to the theatre and musical shows, the circus Created a refined viewing environment and multiple productions. The Strategy Canvas for Cirque du Soleil is shown above. This Blue Ocean Strategy thinking yielded very good results and while other circuses are closing down,

Cirque du Soleil is thriving. The unique offering of Cirque du Soleil can be seen by having a look at its website at:

http://www.cirquedusoleil.com/en/home/shows.aspx

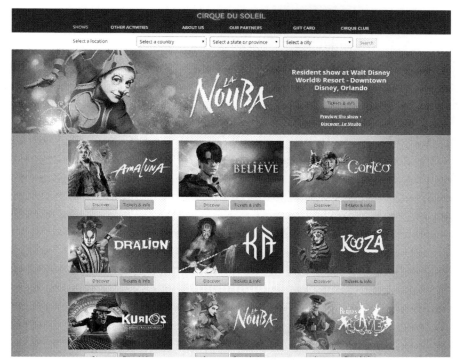

Cirque du Soleil's Website

Now, as an exercise, try using the Blue Ocean Strategy to improve the domestic refrigerator and the process of applying for a credit card.

4.2.5   Trend Recognition

Innovation is driving technology to be better, faster, lighter, smaller, easier to use and cheaper. This drive towards a perfect state is represented by some trends that can be recognised as pathways for technological advancement. Observing these trends, we can use them to generate new ideas for products, services and solutions. It is necessary to mention here that trend recognition is not a methodology to predict the future, but mainly a technique to generate ideas. The reason that I present this caveat is that our track record of predicting the future of technology is less than impressive. I was born in 1970 and since my childhood, I liked to read popular scientific materials. During the seventies, there was an obsession with predicting how the world would look like in the year 2000. The articles and books I read featured robots that did the house chores, flying cars and school trips to the moon. None of the articles predicted the Internet, Google or Facebook. In short the futurists predicted an energy revolution not an information revolution.

Enough with the caveat and let us look into the trends that can be observed in a number of domains. These will be discussed in the following sections.

4.2.5.1 Materials becoming smarter

Most of the materials that we have used since the dawn of civilisation are passive. They do not respond and change with the changes in their environment. This is slowly changing and products are being increasingly made out of adaptive materials that can adapt to the usage requirements and the environment. Understanding this trend can be a useful and straightforward technique to Conceive new products, if you already have a product that is made of a conventional material, ask yourself, can I make it out of an adaptive material? This technique can be used to Conceive products that vary from smart furniture that will adapt to the shape of human body and its temperature to airplane wings that can change with different flying conditions.

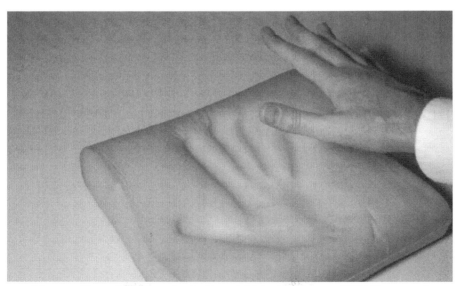

Memory Foam, an Example of a Smart Material (Source: Wiki Commons)

4.2.5.2 Dematerialisation: From Atoms to Bits

Many of the products that used to be physical in nature now have an electronic version or have moved into the electronic realm all together. These include books, music, tickets and receipts. A number of services have moved into the virtual space, these include bookstores and other retail outlets. With the advancement of rapid prototyping technologies and 3D printers, the trend seems to be that more and more products will be sold online where companies will be selling the digital files for variety of products. Customers can then download these digital files and get them printed at home or at local printing centres. So instead of shipping atoms and physical objects, the trend is heading gradually to shipping bits through the Internet.

When searching for ideas for improving products and services, ask yourself this question: Can I make the product or the service digital? Currently, researchers are working on 3D printers that can print variety of materials from metals to plastics and even electronics and food. Imagine being able to download pizza or your own car over the Internet one day!

3D Printing Technology. Enabling the Transformation from Shipping Atoms to Shipping Bits

4.2.5.3 Space Segmentation

The next time you iron your shirt, think of the journey of the iron as a product! It started as a solid block of metal that is heated externally using coal. The improved version was a hollow iron with burning coal inside, this resulted in less material used and longer heating time, since the heating element is carried within the iron itself. Current irons are hollow structures with multiple holes for steam release. This depicts the story of the space segmentation trend. This is briefly shown below:

Solid → Hollow Structure → Structure with Multiple Holes
→ Porous Structure.

The Evolution of the Clothes Iron (Source: Wiki Commons)

So when you have a product that is made of solid material, think "Can I make it hollow?"

4.2.5.4 Surface

Rough surfaces have many advantages over smooth ones. As a matter of fact, under the microscope, no surface is perfectly smooth. Interestingly, even from a drag reduction point of view, it is better to make a surface rough. That is why golf balls have those distinctive dimples. Often making a smooth surface rougher with dimples or ribs will improve it in one way or another. This could be to reduce the drag on an airplane wing, increase the tyre contact with the road, improve the heat dissipation from engines or improve the grip of a coffee cup.

So when you see a smooth surface, explore what benefits will result if it is made rough with dimples, ribs or spikes.

Golf Balls have Dimples to Reduce Drag (Source: Wiki Commons)

4.2.5.5 Object Segmentation Evolution

From a segmentation point of view, objects and products may go through the following evolutional stages:

Solid → Segmented Solid → Flexible → Fluid → Field

It is not necessary that a given product goes through all the stages one by one, but it is clear that there is a shift of technology in this direction. For example, to measure length we use a measuring stick (solid); for larger lengths we developed segmented measuring sticks (segmented solid) and measuring tapes (flexible), and for even larger lengths we developed laser-based devices to measure length (laser as a measuring device is an example of a field).

The Evolution of Measuring Devices (Source: Wiki Commons, GetMeter.com)

4.2.5.6 Size

Most of the devices start their life bulky. As time goes by they become smaller, then they move towards shrinking in size to be at the micro-scale level and eventually in the nanoscale level. This tells the story of almost everything. The first computers needed dedicated buildings to house them, but they now can fit on very small electronic boards within our TV sets, cars and refrigerators. If you want a new idea for a product, you can ask yourself simply, can I make it smaller?

The Evolution of Mobile Phones throughout the Years (Source: Nokia.com)

4.2.5.7 Multiple Functionality

Mobile Phones was only Used to Make Calls and Send Text Messages in the past. Now They Have All Kinds of Functions.

If a product performs one function, it is called a mono-system. If it can perform two or three functions it is called bi-system and tri-system respectively. Those products capable of performing multiple tasks are called poly-systems. An example of mono-system versus poly-system is the ordinary knife versus the Swiss Army knife. This evolution is best demonstrated by observing how mobile phones evolved from devices mainly to make and receive calls to a device that takes pictures, organises appointments, plays music, and an endless list of functions.

4.2.5.8 Price: The Journey to Zero

Often new products are sold for a high price when they are first introduced, and only few people can afford them. As the innovation continues, the prices decrease. Interestingly, with the introduction of the information revolution, many products are not only becoming cheaper but also absolutely free. Think of free e-mail, Facebook, Twitter, newspapers, news, YouTube channels, and all the other products that the internet has made possible. This trend is moving beyond the electronic products where innovative business models are making more and more products free for the consumers. Chris Anderson in his book 'Free' outlined fifty different business models that are built on being free.

Moving towards free can be used to Conceive a variety of business models and products. You can take any current product or service that is sold now and ask yourself these questions:

1. How can I make it free?
2. How can I make part of it free?
3. How can I make it free to part of the consumers?

4.2.5.9 Personalisation

Industrialisation was responsible for making products and services available to the masses at a reasonable cost. However, as humans, we are individualistic creatures, and we always yearn to be treated on that basis. Now the technology is progressively allowing personalisation of products and services, and this trend is expected to continue. I wrote this section on January 7th, 2014, the Google doodle today is of the late Yasmin Ahmad, an award winning Malaysian film director, and today is her 56th birthday. Detecting that I was accessing Google from Malaysia, Google was able to personalise my experience. This trend is now moving into other goods and services as well. Credit cards, for example, can be personalised. While customised cars are still expensive, the trend is clearly moving towards this direction

Google Personalises its User Interface (Source: Google Doodles)

When contemplating or Conceiving new products and services, we can always ask the question "How can I make it more personalised?" This is true if we are producing a car, a credit card or even a haircut.

4.2.6   Biomimicry and Learning from Nature

Nature is a great source of inspiration for everyone and especially for engineers and designers. Knowingly or unknowingly, we have been using nature as a source of inspiration for new ideas. Take for example the old dream of flying. Humans envied birds for millennia and the first attempts at flying were sheer imitations of the birds using wings fitted with feathers. Now endless products that we use and take for granted are inspired by nature. This includes RADAR technology which imitates bats, and high performance swimming suits that are inspired by shark skin. If we imitate nature successfully, we shall be able to build high performance pumps that imitate the heart, powerful computers and software that resemble the brain, filters that copy the principles of kidneys, and high efficiency photovoltaic cells that are inspired by the tree leaves.

Next time we are faced with a challenge we can ask ourselves,

- Has nature provided a similar solution?
- Can we adapt a similar solution to the challenge we are facing?

4.3   Concept Evaluation and Selection

We have thus far explored a number of ideation techniques. Clearly the choice of the technique depends on the nature of the challenge as well as the personal preference. Naturally, ideation techniques yield numerous ideas and concepts and it is essential to have a systematic way to classify these ideas and select the most appropriate ones. After generating a number of concepts, a selection process needs to take place to allow convergence on the most suitable concept and moving into the Design stage. To systematically achieve this, we can do the following:

1. Cluster similar ideas together and integrate them. This will reduce the number of ideas by eliminating repeated ideas and combining ideas that can be integrated together.

2. To converge on the final idea a team can either adopt a voting strategy or use the decision matrix.

When voting is used, just like with the Brainstorming technique, each member of the team is given 3 votes which they can assign to their favourite idea(s). The idea that receives the highest number of votes will be selected.

The decision matrix is a more structured way to select ideas and it requires more work. If we are in the process of CDIO-ing a new racing bicycle for bicycle manufacturers and we have converged on 4 main concepts, the construction of the decision matrix takes place as follows:

1. The team identifies the top 5 features or attributes in a descending order of importance as required by the client. So let us say that these attributes are:

     5: Speed   **4**: Lightweight   **3**: Reliability

     2: Ease of manufacturing   **1**: Cost

Clearly for a racing bicycle, speed is the most important requirement while the cost is the least important.

2. Now construct a table with the row headings referring to the different attributes and the columns referring to the different concepts generated. Each of the 4 main conceived concepts needs to be assigned a number in relation to each of the 5 attributes. These numbers will range from 1 to 4 (the number of the studied attributes). A value of 1 indicates that a concept contributes the least to the given attribute while a value of 4 indicates that a concept contributes the most to the given attribute. So if we take the attribute of speed, the concept that contributes the most to speed is rated 4 (this should equal the number of concepts being

considered) and the concept that ranks the lowest is rated 1. This is repeated for the other concepts. The decision matrix is shown below. The number each concept ranks on a given attribute is multiplied by the weight (importance) of that attribute. Adding these together will give the score of each concept. Although concept 2 did not rank the highest on all the attributes, it is clearly the winner of this comparison.

Decision Matrix

Attribute	Weight	Concept 1	Concept 2	Concept 3	Concept 4
Speed	5	3x5	4x5	2x5	1x5
Light weight	4	4x4	2x4	3x4	1x4
Reliability	3	2x3	4x3	3x3	2x3
Manufacturability	2	2x2	3x2	1x2	4x2
Cost	1	2x1	1x1	3x1	4x1
Score		43	47	39	27

After completing the Conceiving stage, a final test for the quality of the selected concept is necessary. Answering the list of questions below can help ensure the quality and adequacy of the selected concepts at the end of the Conceiving process:

1. Is it desirable? Will the customer desire what I am making? Does it solve a challenge or satisfy a need?

2. Is it economically viable? Will the customer be willing to pay for it? Is there a way to make it cheaper and add more value?

3. Is it feasible? Is there technology to make available?

4. Is it ethical and legal? Does making or selling it infringe on any law or intellectual property (IP)?

5. How can I make it safe for both those who make it and use it?

6. How can I make it easy to make and use?

7. How can I reduce its impact on the environment even when it is no longer in use?

1. In order to conceive ground-breaking, value-adding ideas that are actionable through the design, implement, operate stages of the CDIO process, the following techniques can be used.

   - Brainstorming
   - Random Entry
   - Trimming
   - Blue Ocean Strategy
   - Trends Recognition
   - Biomimicry and Learning from Nature

2. And in order to select the most promising conceived ideas, we can use one of the following methods

   - Voting
   - Decision Matrix

Practical Takeaways

1. Always start the conceiving stage with an open mind and trust the process.

2. At the end of the conceiving process, always use a checklist to ensure that you satisfied all the requirement and constrains.

Chapter 4
Conceive

Chapter 5
Design

Design / dɪˈzʌɪn/
Decide upon the look and functioning of (a building, garment, or other object), by making a detailed drawing of it;
Do or plan (something) with a specific purpose in mind.

Oxford Dictionary

*"**Design** is not just what it looks like and feels like. **Design** is how it works."*

Steve Jobs

Design can mean different things in different contexts, from art to engineering to product design. In general, design refers to the clever arrangement of different components to work harmoniously to deliver a value or perform a task. The Design stage of the CDIO process refers to the cumulative and iterative process of bringing imagination a step closer to manifestation through the use of science, mathematics and common sense to convert resources and achieve prescribed objectives. It is important to stress here that the Conceiving and Designing processes are not separated but integrated and working on the Designing stage may require going back to tweak or even totally change the ideas developed at the Conceiving stage.

5.1   Function and Form

Design refers to creating the marriage between the form (the look) and function of products and objects. Whether what is being designed is a car, a television, a website, a book cover or a survey form, good designs result in products that are functional, safe, easy to use, and economical as well as appealing. Deciding on function and form happens at different system levels. Let us consider designing an office door and let us call it the "super-system." Its subsystems will include the door plank, hinges, doorframe and doorknob. Note that each subsystem may be considered as a super-system in comparison to its components, and can be analysed in a similar way. The function(s) of a super-system are called the "requirements". Often the form of the super-system is a very important part of the requirements as well. When you buy a car, your requirements would include a certain speed and performance, size, accessories as well as look and colour. As we move down into the direction of the subsystem, the importance of form reduces if the parts are less visible and are not in direct contact with the end user.

It is useful here to define the concept of affordance which refers to the property of an object or environment that enables the user to perform a function. The doorknob, for example, affords twisting and enables the function of locking and unlocking the door. Likewise, a slot affords inserting, and a button affords pushing. Good designs ensure that affordance is made visible and mapped clearly towards the desired functions so that the user can easily figure out how the product can be used. More on this is discussed later in the book.

5.2 The Design Process

The Design process is very evolutionary and iterative process by nature. At different stages of the design process we may need to go back to experiments and even to repeat the Conceiving process. The Design process is used to design super-systems as well as subsystems and sub-subsystems. For example, a design team may be designing a car (super-system) while other teams are designing subsystems such as the engine or the exhaust muffler for the car. As the Design evolves, all teams need to align their work and continue communicating with each other to ensure an integrated design that meets the requirements.

Armed with ideas and concepts conceived at the Conceiving stage, generally the design process evolves through four stages, these are System Architecture Design, Configuration Design, Integrated Design and Detailed Design. For easy recall we can use the acronym ACID (Architecture,

Configuration, Integrated and Detailed). These evolutionary stages are outlined below.

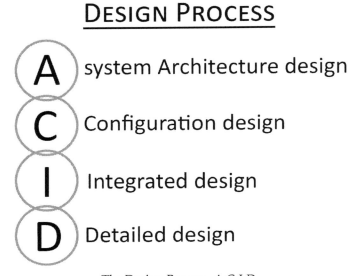

The Design Process. A.C.I.D.

5.2.1 System Architecture

In this stage the main subsystems (components) of the super-system (product or process) necessary to deliver the super-system's requirements are identified. The components are selected based on their functions that, when orchestrated together, will deliver the symphony of the desired requirements. If we are to design a product, let us say a mobile phone, the system architecture design will identify the main subsystems of the mobile phones as:

- Microprocessor
- Microphone and speaker
- Camera
- Touch screen
- Battery
- Operating system
- Printed Circuit Board
- Antenna

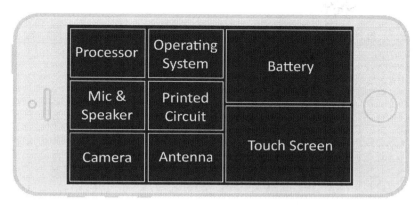

System Configuration Design of a Mobile Phone

System architecture design is also applicable when designing a digital product such as a phone app, software or a website. For example, if the design assignment involves designing an online bookshop, the system architecture will include the different functions and modules involved which may include:

- Search function
- Payment module
- Book recommendation module
- Customer feedback section
- Frequently asked questions

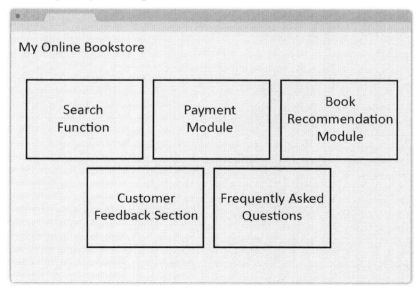

System Configuration Design of a Website

Likewise if we are designing a process to receive customer feedback, the process architecture will involve identifying the main users and the main actions, the service level agreement, and the maximum period of time before the customer will receive feedback. This can include customer submission points and decision-making stages and assessment points.

The outcome of this process is normally a sketch that depicts all the components of the product, system or the process and how they connect with each other. When performing the system configuration design, it is essential to consider the legal and economic requirements governing the use of the final product. For example, in order for cars to be sold in certain markets, it is compulsory for them to have airbags and catalytic converters. Certain energy saving technologies are also becoming part of the requirement for home appliances to be marketed in some countries. This understanding should be reflected when selecting the subsystem components.

5.2.2   Configuration Design

At this stage the product starts to take shape with decisions about the forms and functions being made. These decisions will include modelling, sizing, and materials and components selection. How these decisions are made will be based on the customer's requirements and budget. Final selections will depend on the designer's experience as well as testing, trial and error, and computer simulations. Referring to the mentioned mobile phone example, the configuration design will involve identifying the specifications of the following:

- Mobile phone dimensions
- Mobile phone weight
- Microprocessor speed and manufacturer
- Microphone and speaker capacity
- Camera resolution
- Touch screen size and sensitivity
- Battery size and rating
- Operating system (homemade or third party)

As the configuration design progresses, it is useful to start building soft models of the final product to see how it will look and handle. The picture below shows foam and cardboard models of the famous Polaroid camera. These soft models are cheap to make and they will give the designer an idea of what the final product will look like, allowing quick changes and improvements to be made.

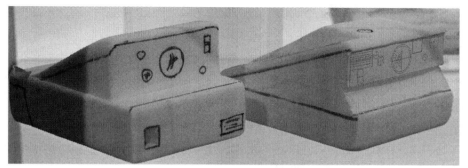

Polaroid Camera Soft Models

5.2.3   Integrated Design

This is a step closer to the final design where systems start to emerge and interactions between different parts are seen. In this stage, the designer ensures that all parts, hardware and software are compatible and are able to fit together cohesively. This process involves modelling, simulation and optimisation of different parts of the super-system, including any software requirements. Considerations such as manufacturability, recyclability, maintenance as well as legal requirements, which include not infringing on patents and intellectual properties of others, are documented and taken care of. Often a rough working prototype made with soft tools and using rapid prototyping technologies emerges from this stage of design and these can be subjected to further testing and fine-tuning.

5.2.4   Detailed Design

Completing all the three steps above, the designers will be in a position to produce detailed designs and drawings that can be shipped across the world to be Implemented and manufactured. Besides these detailed engineering drawings, the detailed design stage also results in a detailed Bill of Materials (BoM), which is a list of all the parts that make a given product. The BoM outlines the names, part numbers and quantities of all the components making the final product. A final cost of the product is also worked out and this will help make decisions on how the product will be manufactured. An example of a simple BoM for a bicycle is given as an example.

Bill of Materials (BoM)

Item	Part Name	Part No	Qty.	Unit Price ($)	Supplier
1	Tyre Rim	BC-100-TR-x01	2	5	ABC Pvt Ltd
2	Tyre	BC-100-TT-01	2	3	XYZ Pvt Ltd
3	Frame	BC-100-FR-01	1	15	In house
4	Saddle	BC-100-SD-01	1	5	In house
5	Bolts	BC-100-BT-01	20	0.2	A-Fasteners
6	Nuts	BC-100-NT-01	20	0.2	A-Fasteners

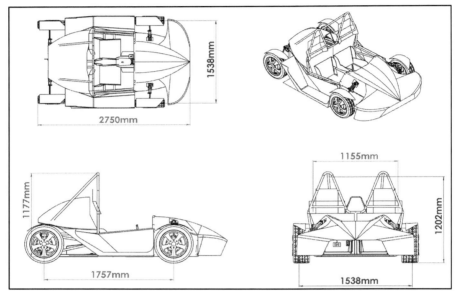

Detailed Engineering Drawing

5.3   Design Optimisation and Trade-Offs

Often designers are faced with conflicting requirements when they perform their tasks. These include making the product both strong and light (maximise strength and minimise weight), durable and affordable (maximise durability and minimise cost), minimise the development cost and minimise the environmental impact and so on. Balancing these conflicting requirements requires trading off some features or capabilities in order to achieve others which are of more importance. This is called design optimisation and it can be achieved using rule of thumb as well as some mathematical and statistical tools.

5.4   Other Design Considerations

It is necessary to mention here that a major part of the design is to predict how a product or a process may fail to deliver the required outcome and attempt to mitigate that. The failure could be a mechanical or material failure, where a product component may breakdown with repeated use or unexpected surge of load. To mitigate this failure, designers need to "over-design" the components that support most of the stresses through adding a Factor of Safety which is defined below.

Factor of Safety - ratio of Design Stress to Normal stress

Design Stress - the stress a component is designed to support before failing

Normal Stress - the stress that a component is normally exposed to

Design redundancy is another way to mitigate component failure and increase product reliability. If the design requires a pump or a filter, and the cost of failure of the pump is very high, a designer may opt to install two pumps instead of one. This way the designer will ensure functionality if one of the pumps fails or is being serviced or replaced.

Other considerations to be taken while designing include:

1. Design for Manufacturability: Designers need to make sure that manufacturing and delivering the product or service is no more complicated that it should be. One way to achieve this is standardisation of similar parts and fasteners. Imagine having a product that you can put together using screws and bolts and nuts of similar size, requiring the use of fewer tools.

2. Design for ease of use and operability: Good products are intuitive and easy to use. Understanding what a product affords and make these affordances visible so that the user knows what to press, or what to turn in order to safely operate the product. Designers also can introduce constraints that make wrong usage of the product not possible. If the batteries need to be inserted in a certain way into an electronic appliance, the manufacturers often put a sign indicating the correct way of inserting them. A better design would be to make it impossible to insert the batteries incorrectly.

3. Design for safety: Designers need to ensure that they take the safety of the users into their consideration. If the product is to be used by small children, then neither the product nor its packaging should contain hazardous parts such as small parts that children may insert in their ears or noses, or any parts that represent a choking hazard. Other safety considerations would be the avoidance of toxic materials and materials that are a fire hazard.

4. Design for aesthetics: We mentioned earlier that function and form are both important features of design. In this era of abundance, building products that simply do the job is not good enough. Products need to look and feel good as well.

5. Design for sustainability: With all the environmental challenges that we are facing in this century, designers and engineers have only one choice, to design for a sustainable future. This is done through making products out of less materials and making them to consume less energy during both manufacturing and operation. The materials used need to be recyclable and the design itself needs to make the process of recycling easy.

6. Design for affordability: If the product is designed to be introduced to the masses, then designers need to make decisions related to the materials used and the manufacturing processes, and even functions to ensure that the price of the product is within the reach of the market segment being addressed.

At the end of a good design stage, you will need to be able to check off most of the points below in relation to the product or service you are designing. This includes:

☑ The design is desirable. It looks and feels good.

☑ The design is technologically feasible. The technology to make it is available.

☑ The design is economically viable. Users will be willing to buy for the recommended price.

☑ The design is legal and ethical. It does not infringe on any law or IP

☑ The design is environmentally sustainable.

☑ The design is easy to manufacture

☑ The designed system is easy to operate and maintain

☑ Five ways the system may fail have been identified and mitigated in the design.

☑ Five ways the user may use the system wrongly and/or unsafely have been identified and measures to prevent them have been incorporated in the design.

☑ The engineering drawings and/or the software outline are clear.

☑ The BoM is exhaustive and clearly documented.

Highlights

1. The design process involves the following stages (ACID):

   - A - System Arcitecture Design
   - C - Configuration Design
   - I - Integrated Design
   - D - Detailed Design

Practical Takeaways

1. Trust the process and ensure that you systematically go through all the steps outlined above.

2. At the end of the designing process, always use a checklist to ensure that you satisfied all the requirement and constrains.

Chapter 5

Design

Chapter 6
Implement

Implement / 'ɪmplɪmɛnt /
Put (a decision, plan, agreement, etc.) into effect

Oxford Dictionary

After a product, solution or process has been Conceived and Designed, it is time to execute it and bring it to life. This stage of the process is called Implementation and it is as important as the other steps in the CDIO cycle. Many good plans and designs fail to achieve their objectives simply because of poor Implementation and execution. The Implementation stage involves a great deal of management and coordination of various resources including time, human, and physical resources. It also requires integrating software and hardware components, working with various multidisciplinary teams and the ability to communicate and manage projects well.

6.1 Hardware Manufacturing Process

After the design is completed, both the hardware and software (if any) need to be implemented. Depending on the volume of the product to be produced as well as the available budget and timeframe, a suitable manufacturing technique is selected. This process will start with producing a working prototype which can be used to perform further testing of the product. The prototype can be constructed using a variety of manufacturing techniques including the use of CNC machines, rapid prototyping machines and the use of soft tooling and vacuum forming. The prototype serves as a proof of concept for the suitability of the selected manufacturing process.

Taylor's Racing Car. From the Drawing Board to the Racetrack.

6.2 Software Implementing Process

The software content of a given product varies from zero, for basic hardware products, to almost 100% for software-based products such as websites and computer-based systems. Whatever the case, the software needs to be programmed using a suitable programming language, making sure that the factors and units used in both the hardware and software are homogenous. This is especially important when working in multinational teams or with multi-sites where different systems of units are used.

iPad controlled Robot Project by Foundation in Engineering Students

6.3 Hardware Software Integration

More and more products and appliances have software intelligence in them. This is true for cars, televisions, and even home appliances such as microwave ovens and refrigerators. Software is necessary to control and optimise the performance of many of the products that we use today ensuring low energy consumption and providing connectivity and convenience. For the successful and safe use of a product, the hardware and software should integrate smoothly. NASA engineers learned the importance of proper hardware-software integration the hard way when they lost the Mars Climate Orbiter on September 23rd, 1999. The orbiter was Conceived, Designed, Implemented and Operated by NASA to study the climate of Mars. The orbiter (hardware) was designed based on the SI (metric) units while the ground-based computer software produced output in Imperial units. This resulted in the sending of incorrect instructions to the orbiter causing it to come dangerously close to the Martian atmosphere and disintegrating. The orbiter cost USD125 million to develop and launch.

6.4 Testing, Verification, Validation, and Certification

The final system, service or product with hardware and software fully integrated needs to be tested to verify and validate that it meets the design requirements. Systems, services and products can be exposed to a barrage of tests depending on their nature and application.

Engineers need to develop a "Testing Plan" to verify the items that need to be tested and validate that these tested items conform to the design specifications and requirements. The Testing Plan is a document that outlines the testing protocol indicating the test parameters based on the design specifications of the system or product, the testing range, the type of tests and the tools for the tests. The protocol is aimed at validating the achievement of the customer's requirements as represented by the design specifications. The tests can be for the super-system and each and every one of the subsystems. Many of the products that involve the safety of people, such as cars, children's toys and home appliances, need to be tested and certified by independent bodies before they can be introduced into the market.

These tests can include the following:

1. Dimensional test: To verify that the product conforms to the designed dimensions.

2. Visual test: To verify that the colours, look and feel are according to the design specifications.

3. Functionality test: To verify that the system performs the functions it is designed to perform.

4. Reliability test: Products need to ensure that they can perform their functions over an extended period of time and in variety of environmental conditions.

5. Power measurements: If the product has a power rating, this needs to be tested and verified.

6. Signal quality test: For products that have some electronic signal exchanged, it is necessary to test for the quality, frequency, interference and strength of the signals.

7. Safety tests: Some products such as cars require destructive (crash) tests before they are certified as roadworthy. These tests can be done using computer and physical models. Some other products need to pass variety of stress tests before they are certified.

Partial Checklist for a Racing Team Test

RACING TEAM DESIGN TEST CHECKLIST	
Chassis Tests (Digital)	Pass/Fail
Roll Protection Compliance Test	Pass
Static Chassis Stress Test	Pass
Turning G Stress Test	Pass
Torsional Rigidity Test	Pass
Frontal Impact Test	Fail

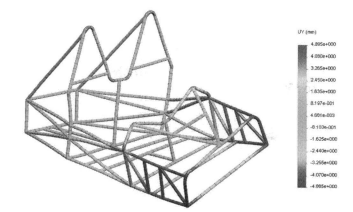

Software Testing of Torsional Rigidity for the Race Car Chassis

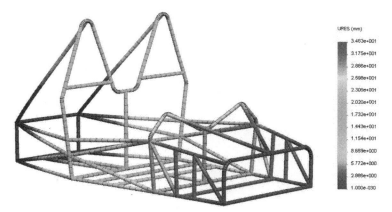

Software Testing of Side Impact for the Race Car Chassis

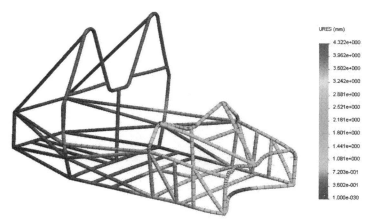

Software Testing of Front Impact for the Race Car Chassis

Highlights

Never directly jump to the implementation stage without going through the conceive and design stages

Chapter 6
Implement

Chapter 7
Operate

Operate / ˈɒpəreɪt / v.
Manage, work, control; put or keep in functional order.

Oxford Dictionary

To complete the CDIO process, engineers need to ensure that whatever they have Conceived, Designed and Implemented is safely, effectively and sustainably Operated. They also need to ensure when the product or system they have created reaches the end of its life, it can be retired with the least impact on the environment.

7.1   Sustainable and Safe Operations

The Operation stage of the CDIO process can be the longest in the entire CDIO process. A bridge, for example, can take few years to construct but can remain in Operation for centuries. Likewise an airplane or a car can remain in service for decades. During the Operation stage, engineers need to ensure safety and sustainability, both economically and environmentally.

7.2   Operations Management

Depending on the complexity of the system, Operation management includes installation, commissioning, testing, maintenance and daily operation. If the system is a petrochemical plant, commissioning and testing could take up to weeks or even months to ensure that all the parts of the plant are working well and that all the safety equipment is functional in a variety of operational scenarios.

7.3   Training and Operations

Whether you buy a new car, mobile phone, television set, software package or furniture from IKEA, they all come with a user manual or user guide that outlines the best way of using them and how to take care of them. In the case of IKEA furniture, the user guide shows how the furniture is assembled. For more complex systems, manufacturers need to develop maintenance manuals and training materials for the service and maintenance staff. This includes a certification programme for the maintenance staff. Examples of highly complex systems would include airplanes that are sold to airlines around the world. Manufacturers of airplanes run training and certification programmes so that the technical staff operating and maintaining the planes are able to perform their intended duties.

7.4   The Birth of the Checklist

The Boeing Model 299 (later renamed B17 Flying Fortress) was one of the most impressive long-range bombers of its time. The plane had a wingspan of 32 m and a length of 23 m. Equipped with 4 engines; the plane was fast, had a great range and was able to carry a tremendous load.

October 30, 1935 was the date chosen for the plane to be tested at the Wright Air Field at Dayton Ohio by Boeing's chief test pilot, Leslie Tower, and Major Ployer Peter Hill, the air corps' chief of flight-testing. The plane taxied down the runway and after reaching an elevation of around 100 m, the plane stalled and crashed into the ground killing both pilots. The air crash investigation concluded that the cause of the crash was 'human error.' The B17's might came at a price, it was very complex to operate. Every one of its four engines had its own indicators that needed to be monitored individually to maintain stable flight conditions, which required pilot intervention. Pilots also needed to monitor and attend to a large number flight conditions. While doing all this, the pilots forgot to release a newly introduced locking mechanism on the elevator and rudder controls. And this single mistake doomed the plane.

The army was so impressed with the design, that despite the tragic accident, it wanted to buy the plane. The challenge was how to safely fly such an inherently complex four-engine plane? The solution that the manufacturers came up with was both simple and innovative! First of all, no one suggested getting the pilots to undergo more training. The pilots who made the mistake were simply the best in business. Instead, the plane manufacturer engineered checklists for the pilots to consult and go through prior to flying the plane and at each critical flying stage. Believe it or not, prior to the B17 crash, pilots learned the systems of aircrafts and operated them from memory! This was manageable when the list of things to remember hovered around half a dozen. However, when the list of things to remember and execute was too long, human memory is challenged and the pilots were finding it easy to miss simple (but essential) steps. The checklist for the B-17 is shown below.

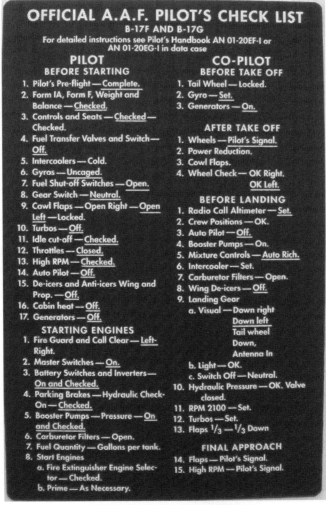

Official B17 Pilot's Checklist

The Crashed Boeing Model 299 (Source: USAF National Museum)

Today, every plane, no matter how small, features some kind of a checklist. Bigger planes have books worth of checklists to assist the pilots with various scenarios in every flight phase.

As a matter of fact, nowadays medical practice has taken the initiative to use checklists. Surgeons around the world are using checklists to ensure that they are operating on the right person, performing the correct procedure, and making sure that various patient specific precautions has been taken into consideration. When I accompanied my wife in the operation theatre while she went through a surgery to give birth to our third son, I observed the surgeon calling for "time-out" before he made the incision. During this "time-out", the surgeon went through a checklist with his team to ensure that they are operating on the right patient and that all the necessary precautions in relation to my wife's condition were taken. The implementation of checklists in healthcare has now a proven record of improving the outcome of various medical procedures.

Whenever we operate a system, it is useful to create a checklist that will ensure that different users are able to safely and accurately operate the system. My students learnt this the hard way when they built a racing car and shipped it outstation to be raced. Relying on memory, they were confident that they packed everything they needed. They had multiple sets of spare parts, engine oil, and they even brought their own welding machine and generator just in case they need to repair the car. To our horror, we discovered that we had forgotten to bring the fire extinguisher, which is required by the race regulations to be installed in the car. Luckily, the students managed to quickly drive to a nearby town and purchase a new fire extinguisher. Learning from this incident, my students now have a number of checklists to help them manage the main phases of operating a racing car.

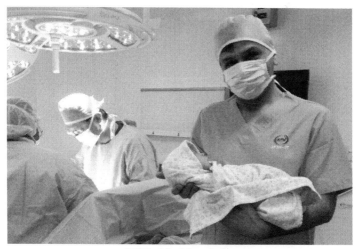

With My New-born

CHECKLIST FOR TOOL AND ITEMS (RACE & TRACK TESTING)

Items	Check Box	Items	Check Box
EE Department		Duct Tape***	
Multi-meter**		Carburettor jet box**	
Long Nose		Extra Carburettor**	
Wire Cutter		Marker Pen	
Screw Drivers		Chain Lube	
Wires		Wire Lock (Steel Wire)	
Black Tape		Used Battery	
Fuse (10A,30A)**		WD-40	
Cable Tie***		Engine Oil	
Soldering Tool**		Brake Fluid	
Jumper Wires**		Aluminium Block	
Switches		Measuring Tape	
Batteries*** (12V & 9V)		Solid Rod*	
Shrink Tubes		Tubing (Small & Big)*	
Lighter		Steel Plates*	
Connectors		String (Alignment)	
CDI (Epifany & Imperica)**		Cable Ties	
Brake Lamp (Backup)		White Board (Lap Time)	
Crumble Wire		Petrol Tong	
Tachometer (Backup)			
Mechanical Tools			
JTC Tool Box (Full Set)**			
Allen Keyset (Full Set)**			

TRT Pre-departure Checklist (Partial)

7.5   Preventive Maintenance

The cost of system breakdown is often high, resulting in down time and economical losses. If the system is an airplane, for example, a breakdown or a failure while airborne could end with catastrophic results. Preventive maintenance is the scheduled services and/or replacement of different parts of the system well before their scheduled breakdown or failure. Preventive maintenance can result in smooth operation as well as economical savings through scheduling maintenance and service activities during the off-peak seasons and ensuring optimum operating conditions during the peak season.

7.6   System Improvement and Evolution

Often complex systems such as dams, buildings, bridges, and airplanes are designed and implemented keeping in mind the possibility of future evolution and improvement. For example, there is always a possibility of upgrading the engines of the airplanes or the addition of new control equipment for a dam to make them more efficient. Infrastructure systems, such as sewage and road systems require continuous improvement and evolution to deal with the changing requirements and volume of use.

7.7   End-of-Life Issues

Many products and systems have in them materials that are toxic, harmful to human health and the environment, and even precious materials. It is necessary to plan what will happen to these materials when the product or system is no longer in use. This planning needs to take place during the Conceiving and Designing stages. The use of hazardous and difficult to recycle materials should be reduced to mitigate end-of-life issues.

7.8   CDIO Case Studies

The School of Engineering at Taylor's University adopts the CDIO framework as an educational philosophy. Starting from the first semester of their studies, students work in groups on major projects to Conceive, Design, Implement and Operate engineering systems. The CDIO process and the project-based learning are aimed at empowering the students to pursue their life's purpose and achieve their full potential while working on projects that are challenging, interesting and meaningful. Throughout the years, students and lecturers worked together on a variety of projects. We succeeded at times and failed at others, but there was always valuable learning and discovery. Often the most valuable discovery is learning more about one's own self and one's potential. This elucidates the point that the CDIO process, when implemented effectively, can provide the backdrop against which human development and empowerment can be realised. Below are some of the stories of the CDIO in action.

7.8.1 Taylor's Racing Team

Taylor's Racing Team is comprised of students from different levels (year 1 to year 4) and different engineering disciplines, namely chemical, electrical and mechanical engineering. The team designs, builds, and races cars at the national and international student races. The spark of the team started in 2010 when a group of first year students wanted to build a car and race it at the Formula Varsity national race in Malaysia. Back then it sounded like an impossible dream, as the team neither had the experience to build cars nor the money to buy all the parts. The team made history by winning the race beating teams from other universities who dominated the race for years.

Taylor's Racing Team in 2010

First the team needed to get someone to sponsor them, after trying more than 100 companies, Red Bull said "yes!" and became the main sponsor. With the money secured, the students built a car and shipped it to the racing track in Melaka, a city 150 km south of our university. The race took place over a weekend and the team performed very well in all the pre-race tests such as the breaking and acceleration tests. During the qualifying race, however, the driver pushed the car over its limit and blew up the engine! Faced with the potential of leaving the race, the team decided to remove the engine, drive back to the university and overhaul the engine overnight then drive back to Melaka to install the engine and get the car ready by next morning just in time for the race. Because the team was unable to complete the qualifying race, the car started in the 13th place on the grid. One by one, we watched our racing car overtaking the competitors and end up being the Grand Prize winner of the race. This was one of the really amazing moments for all of us.

The damaged engine block

Taylor's Racing Team Crossing the Finish Line at Formula Varsity Race 2010

"I always liked cars, as a matter of fact that was one of the reasons why I joined mechanical engineering," Sunny Lee, the team leader said. "However,

I never imagined that my CDIO journey would take me places with Taylor's Racing Team," he added. Now a postgraduate student pursuing his Master's degree, Sunny credits many of his successes to applying the CDIO process. He told me that sometimes the CDIO process might sound a bit tedious and slow as the students always have the tendency to rush into the Implementation and Operation stages, "however almost always when we rush through the Conceive and Design stages, something happens and we regret it later. For example, in 2012, we were racing in a national race, we ignored the design of our cooling system and simply attached a duct to direct the air into the engine. Worse still, we fixed the duct with simple duct tape! Our car was in the first place and during the 23rd lap, when we noticed some smoke coming out of the exhaust. Shortly after that, the car just stopped and we were out of the race." Sunny said. It turned out that the vibration experienced by the air duct was so severe that it caused the duct tape to disintegrate leaving the engine to be "fried" without any air to cool it down.

Imperica's Engine Cooling Ducts

On a more personal level, Sunny attributes the development of his organisational and communication skills to the projects he did while with the Taylor's Racing Team. Being educated in Chinese schools before the tertiary level, Sunny's English was limited when he first joined the university. Working in diverse teams provided him the opportunity and the motivation to speak and write using better English. Recently, he received the Best Student Award from the Institution of Mechanical Engineers (UK) and I was delighted to see him delivering an eloquent speech, crediting his lecturers, peers and the CDIO process!

Sunny Lee

Mike Ooi, another Taylor's Racing Team member, identified the ability to manage conflict as one of his major areas of growth that took place during his work with the racing team. "As an Asian student, I grew up in a culture that has always ignored and avoided conflict. Needless to say, ignoring the conflict does not make it magically disappear!" Mike said. "Being a part of Taylor's Racing Team taught me to speak out and accept that, when working on a complex project with many people of different personalities and backgrounds, conflict is inevitable. While learning the CDIO process we were exposed to techniques that helped me deal with conflict effectively" he added.

Mike Ooi

Justin Moo is the current team leader and driver of the Taylor's Racing Team. He asserted that being a part of Taylor's Racing Team is one of the highlights of his student life. "We have the legacy of the achievement of our seniors to build on and protect," he said. "Although being part of the racing team is very demanding as we are expected to juggle our academic deadlines as well as those of the races, I am proud to say that everyone on the racing team is doing exceptionally well academically with four of us on the Dean's List!" He added proudly.

Justin Moo

Maitha (from Oman), Mayko (from Myanmar) and Andrea Kraal (from Malaysia) are our racing ladies. Being a part of the racing team they proved that engineering is for everybody. "The only drawback is that I am not able to colour my finger nails since I joined the team" Maitha jokingly told me. In general our female students are in agreement that studying in a CDIO environment is preparing them well for their future careers.

The Ladies in Action

Taylor's Racing Team is still thriving with many students aspiring to join it. The students have built three cars thus far and they continue to improve their technology, raise funds, and most importantly grow and develop while evolving their leadership styles. The team website is:

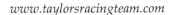

www.taylorsracingteam.com

Taylor's Racing Team in 2014

7.8.2 Women in Engineering

A lot has been said about the lack of women engineering. This subject has been studied from academic, cultural as well as corporate angles. Still, we do not have enough women who are choosing engineering as a field of study and choosing to pursue careers as engineers. Personally, I feel that most of the reasons outlined in the literature to explain why women are not sufficiently represented in engineering, such as gender pay disparity, the hectic life style and the perceived lack of career advancement are simply not unique to engineering. One aspect I believe that we can address in engineering education is the cultural dimension by helping the creation of more female engineers as successful role models.

In 2011 Leena Gade became the first female race engineer to win the world's greatest sports car race, the Le Mans 24 Hours. She repeated the feat again in 2012 as part of the Audi Sport Team and was named the FIA World Endurance Championship's "Man of the Year". Leena has a degree in Aerospace Engineering from the University of Manchester. We definitely need more engineers like Leena so that we can break the stereotype that successful engineers are only males. Showing that engineering can be a highly fulfilling and exciting career option for female students can go a long way in encouraging more of them to take the science and engineering route.

Leena Gade (Source: Huffington Post)

In preparation for this section of the book, I interviewed Yvonne Lim, our top engineering graduate in 2013. She chose to study engineering because she wanted something academically challenging. "I never expected the days and nights worth of projects coming my way. My experience with project-based learning has been a love-hate relationship from the very beginning." Yvonne added.

She went on saying "being one who is blessed academically, projects have been a step out of the comfort zone for me. Unlike exam-based assessments where I have full control over the grades that I can achieve, project-based learning stretched me in ways that I never thought possible. Each semester brought about a new project, new experiences and new challenges.

"Whether it was from the project itself or while working with people, I've never had a challenge-free project. There were times when I had thought I'd be better off without projects. However, I don't regret being through a project-based learning program."

Yvonne said that CDIO helped her as a female student to find her bearings in the engineering world. She felt that the fact, in a CDIO environment, she was able to use the tools and machineries in the workshop, broke some of the barriers in her mind.

"I have learned so much working on projects, from the first project in my very first semester to the capstone project in Year 3 and then the final year project. Projects taught me not only about the theoretical knowledge required by engineers, but also working in teams, managing conflicts, presenting in front of an audience, and most of all working with a deadline. It has also brought me to do things and brought me to places where I had never imagined myself at that point of time. Although projects-based learning came with a lot of stress and sleepless nights, in the long run it has truly been a blessing in disguise. The experiences I have gained, the lessons I have learned, the people I have met and the life-long friends I have made along the way are priceless." She added.

Yvonne is currently doing her PhD in Engineering exploring new ways to store energy as liquid air.

Yvonne with her classmate, Ralina at an engineering conference at MIT

7.8.3   CDIO Beyond Engineering

Christopher Chew is one of my most memorable students. Slim, soft-spoken and with a nice smile always on his face, Chris is a very confident and reliable young man. Describing himself, Chris said "Growing up in my family was such a bliss, with my parents giving me everything I wanted. Essentially I became a spoilt brat, throwing tantrums when the slightest things don't go according to how I wanted. When I joined engineering, I still had a little spoilt brat in me, always feeling entitled. I learnt two major lessons in maturity, the first, you have to work hard to get what you deserve and the second, you need to have an open mind. Both these lessons were taught to me by four "magical" letters; CDIO.

"Naturally, I'm a stubborn person, always thinking I'm right and my methods are the only way to get things done. When I first did engineering projects, I would always jump towards the end, the product, without ever thinking through the process. In my second year, there was an instance whereby my project was delayed and there was insufficient time left to complete it; that was when I took a leap of faith and applied CDIO from start to finish and I was glad to say that it was a success. This was also my second lesson, to get great results one cannot cut corners and take shortcuts. Thinking I was smart enough to skip the C-D process and jump straight to I-O was my downfall on a couple of occasions.

Upon graduation with a Bachelor Degree in Chemical Engineering, Chris opted not to practice engineering and he currently works for Accenture as a technology consultant. He landed this job after a week-long interview process where he competed with other graduates with degrees in business and management. As he consults for some of the biggest companies in the world, he still utilises the CDIO framework to unlock and deliver value. "Now that I'm a CDIO convert, I subconsciously have this systematic thinking in my mind that helps me in everything I do. For example, even when I'm organising an event, I would go through the CDIO process, albeit in a modified for the situation. In an event management environment, I first start off by getting an idea and start to envision the event; that would be the 'C' stage. Then I'll move on to planning, the 'D'. And just before execution, the 'O', I would always set out running sheets or task lists, the 'I'. This form of structured thinking has definitely helped particularly in the areas of time management and goal achievement. Needless to say, I've greatly benefited from applying CDIO and I am now a strong advocate for it." Chris added.

Christopher Chew

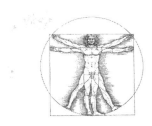

Chapter 8
Ergonomics: Human-Centred Design

"Human-centred design (is) meeting people where they are and really taking their needs and feedback into account. When you let people participate in the design process, you find that they often have ingenious ideas about what would really help them. And it's not a onetime thing; it's an iterative process."

Melinda Gates, Gates Foundation

"Others approach a challenge from the point of view that says, 'We have the smartest people in the world; therefore, we can think this through.' We approach it from the point of view that the answer is out there, hidden in plain sight, so let's go observe human behaviour and see where the opportunities are."

David Kelly, IDEO

"Human-Centred Design helps us identify unvoiced needs. It helps us identify that friction between the user and the world and the way that the user's way of thinking, their mental model doesn't match up at all with the way that the product or experience or software looks like and help reveal those unmet needs."

Mickey McManus, MAYA

Human-Centred Design (HCD) refers to the design process that is informed by human biology and psychology and is aimed at creating systems, products, experiences and processes that are not only easy and safe to use but also create a remarkable and enriching experience for the users. The realm of human-centred design extends from everyday objects such as stoves and cars to control rooms in nuclear plants or airplane cockpits.

8.1 Design for Ease of Use and Operation

Don Norman, in his book 'Design of Everyday Things,' talks about the concepts of "knowledge in the head" and "knowledge in the world." Knowledge in the head refers to the amount of learning that we need to have in our heads prior to being able to operate a given product or system. Norman argues that a good design can result in products and systems that are intuitive and easy to use by making the knowledge needed to operate these products or systems easily available in them and their surroundings - knowledge in the world. We do not need to learn anything about a well-designed door, for example, before being able to open it! The knob affords turning and turning the knob will result in opening it. However, we all can recall examples of poorly designed doors that we need to "learn" how to use! They include doors that have handles that afford pushing and pulling while they are sliding doors and which are supposed to slide sideways. People often end up putting many signs on the door to help the user learn how to use them. Although it is easy to take good design for granted, with so many poorly designed products around us it is clear that the human-centred design process is not automatic but requires meticulous and intentional planning.

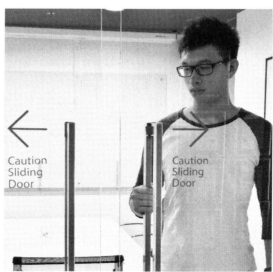

A Door That Affords Pushing/Pulling but its Operation Require Sliding

8.1.1 Affordance, Visibility and Feedback

As mentioned earlier, affordance is a property of objects, environments and media that refers to what they can "afford" or what can be done with them. This is an important design concept when developing systems, products, and services that are easy to use. For example, a doorknob affords twisting, a slot affords inserting, a button affords pushing, a board affords writing on and an Internet hyperlink affords clicking. When the affordance and function are aligned, we generally have good designs. As a rule of thumb, whenever you see the sign "Do not ...." You can expect some failure of alignment between affordance and function. Try to recall the time when you have seen a sign proclaiming "Do not sit!" Often this is related to an object that you can sit on, the object is almost inviting you to sit on it; however, the object may be a box that is too frail and can collapse under your weight. Other examples include, "Do not push!", "Do not pull!", and "Do not write!"

Good designs have a clear link between their affordances and functions. A door handle may afford both pushing and pulling, while the door function "opening" may be achieved through only pulling or pushing. This represents a disconnect between the object's affordance and its function, and thus represents an area for improvement of the design. During the design process, the connections between affordances and functions need to be fully explored to ensure the development of user-friendly objects that can be operated conveniently and with minimum learning or instructions.

An object should advertise its affordance(s), making them visible and inviting the user to perform the right operation in order to safely and correctly deliver the object's function. Affordance can also invite unexpected functions and operations as well. My office has floor-to-ceiling glass panes, besides serving as walls, these glass panes afford being written on, and I immediately used them as writing boards. They served the purpose very well apart from the occasional "loss of data" if our diligent cleaners insist on cleaning my wall. Because flat surfaces "afford" being written on, this at times leads to graffiti and vandalism. Hence if you really do not want people to write on a surface, design that surface in a way that the affordance is less obvious or less inviting.

To close the functionality loop, it is often useful if the product gives feedback after the user performs an operation. If we turn on the iron, for example, a red light is turned on indicating that the operation was successful.

The feedback can be delivered through visual cues, sound or smell. The feedback can be designed to indicate unsafe operation as well. Most cars are fitted with reverse sensors, as we reverse the car, the sensor gives an audible signal that increases in pitch or frequency as the car gets closer to an obstacle behind it. An example of using smells for feedback is the addition of foul-smelling chemical to liquefied petroleum gas (LPG) used in many households for cooking purposes, to indicate if there is a leak. LPG is odourless and without the additive, a leakage can result in nasty accidents.

I shall give a few more examples from my own personal experience. I do not know about you, but whenever in a restaurant, I am not sure which shaker contains the salt and which one contains the pepper (I know it has to do with the number of holes on the top of the shaker, but I could never figure out which is which). I saw some shakers that have the words salt and pepper written respectively on them. A clever solution I saw were shakers with holes on the cover in the pattern of S or P indicating salt and pepper respectively.

Every day when I park my car at the car park at work, I take my bag and lock the car using the button on my key. When the car is locked, it gives a visual feedback, its lights turn on and off three times. However, often after walking to my office, I ask myself "did I lock the car?" It would have been really useful if the car key is able to give feedback indicating whether my car is locked or not.

A Concept of a Car Key with Additional Feature Showing Whether the Car is Locked or Not

Often after ironing our clothes, changing, and leaving home, my wife and I ask each other this question "did you turn off the iron?" In the absence of a confident answer, we often decide to drive back to check. We once forgot that we left the iron on and left home. The iron burned the ironing board and we were lucky that we did not have a major fire. It would have been very useful if we could get feedback on the state of our iron (on or off) through a mobile app for example. It would be even better if the iron is designed so that it turns itself off if there is no activity associated with it for 2 minutes. This will be discussed further in the following section.

8.1.2 Constraints

We are surrounded by thousands of systems, products and services that we need to use on a daily basis. These range from phones and microwave ovens to cars and ATM machines, and there are many ways to operate these products and services incorrectly; hence instructions are often given to the users to enable them to correctly and safely use these products, services and systems. Using constraints refers to designing the product or service in such a way that makes the wrong or unsafe use highly unlikely or even impossible. For example, if batteries are needed to operate an electronic product and the batteries need to be inserted in a certain way, many products have + and − signs indicating where the positive and negative ends of the batteries should be inserted. A better design will be to make it impossible for the batteries to be inserted incorrectly or improve the electronic design so that the device will work regardless of the direction of the batteries.

The first ATM machines used to be operated in the following sequence:

1. Insert the card in the ATM
2. Key in the PIN and details of the transaction
3. Take the money
4. Retrieve the card

However, many users started to leave the ATM machine right after getting the money, resulting in many cases of card loss which overwhelmed the banks. Now all the ATMs around the world are designed with a constraint that forces the customer to retrieve the card before being able to take the money.

Whenever the user is about to perform an irreversible operation, it is very useful to issue a reminder before allowing the action to take place. If we are about to delete a file permanently, computers nowadays ask us to confirm that this is indeed the required task.

Sometimes the constraints are purposely used to restrict an action by the user. I was faced with this when I tried to close my Facebook account. I was looking for an icon or a link that is denoted with "quit" or "close account." To my surprise the process was very obscure and non-intuitive. I concluded that Facebook made it purposefully difficult for its subscribers to quit.

8.1.3 Mapping

Mapping is another way of putting knowledge in the world (rather than the head). It represents a mental model that signifies the causal relationship between actions and their results. This concept is useful when designing control panels and switches. When I teach in our bigger lecture theatres, I sometimes need to turn off the lights at the screen's section if I am playing a video. I often end up with a trial and error process, turning off the lights at the wrong section of the lecture theatre before reaching the switches for the right lights. This situation can be improved if the switches were clearly mapped representing their actual position in the lecture theatre. The same can be said when designing the control panels for a power plant, a chemical process plant, a nuclear plant or a cockpit of an airplane.

A coffee machine at our office has a number of buttons that can be pushed to get different types of coffee, milk and steam. It has three different nozzles to dispense the different types of drinks. The challenge is that the buttons are not mapped to the corresponding nozzles and it is rather easy to place the cup under the wrong nozzle. One of the office staff created additional instructions to accompany the coffee machine to show people how to operate it. When this did not work, the staff created additional labels and stuck them on the machine denoting different actions. The coffee sure tastes great, but only if you can operate the machine. It would have been much easier if the buttons to produce a certain type of coffee are located next to the nozzle that dispenses it!

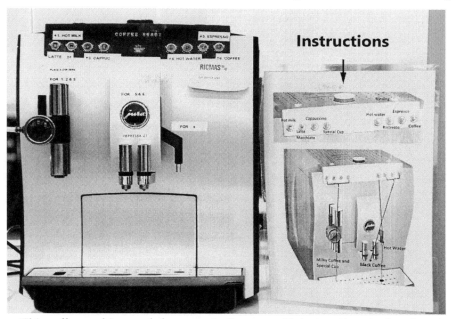

This coffee machine needed a separate diagram with instructions to show how it works. Can we better map this coffee machine?

The image below shows another electrical feature at our university. It represents three power plugs imbedded in the table in a meeting room. It would have been much easier to use if the button was located next to the corresponding socket. Are there any other ways to better map the electrical sockets below?

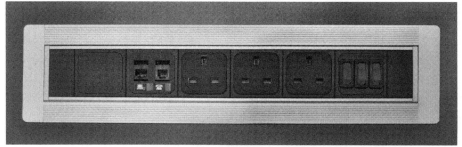

Switches are poorly mapped to their corresponding sockets

8.2   Anthropometric Measurements

Anthropometry refers to the measurement of individuals. It involves identifying standard measurements and variations of the human body such as height, weight and different body proportions. Anthropometry is essential in the design of clothes, cars, doors and many other products that people use.

Anthropometry also looks into the variance in measurement among different ethnic groups. A two-year study conducted by the Indian Council of Medical Research measured the penises of over 1,200 volunteers from different parts of India and concluded that about 60% of the volunteers have penises which are between three and five centimetres shorter than international standards used in condom manufacturing. India has one of the highest number of HIV infections in the world. To curb the spread of sexually transmitted diseases, the government distributed free condoms. However, many men were reluctant to use these free condoms. Baffled by this behaviour, the government commissioned the study mentioned earlier to understand it. The verdict is that because of the large sizes of the condoms, the men were embarrassed to use them. Once understood, this was easily fixed by the condom manufacturers making condoms with sizes that suited the addressed population.

8.3   Work Musculo-Skeletal Disorders (WMSDs)

Work Musculo-Skeletal Disorders refer to the group of painful disorders that are caused by injuries to muscles, nerves and tendons. These disorders normally result from repetitive motions and strains, as well as poor body postures while performing work-related activities or from using a certain product for prolonged periods.

Besides individual suffering, WMSDs come at a considerable cost to the healthcare system, not to mention the productivity loss. To prevent WMSDs, it is essential that work tasks, tools and equipment, and work spaces be designed ergonomically in a way that reduces the strains on different parts of the body.

8.4   Cognitive Ergonomics

Cognitive ergonomics is related to mental processes such as thinking, perception, and memory and how they impact the interaction of humans with different parts of a product or system. This can be useful in designing effective human-machine interfaces and human-computer interfaces and training programmes, as well as effective policies and marketing programmes.

8.4.1 Nudges

When governmental agencies and employers are interested in encouraging employees to save a portion of their salaries, research showed that it is much more effective to give the employees a form that asks them to opt out of the saving programme if they are not interested, rather than opting in if they are interested. This exploits a human thinking feature that is not keen on changing what is established. This is also the reason why when you fill in an online form, the form will often ask to tick a box if you do NOT want to receive regular promotional emails.

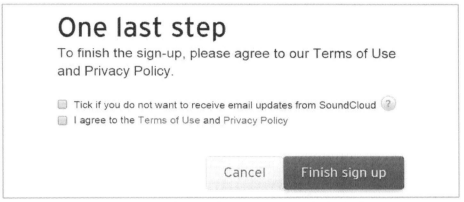

A Screenshot Showing a Website Asking Users to "tick if you do not want to receive more information"

Similar nudges are now being suggested to encourage organ donation. The suggested programme puts all the citizens, automatically, on the list of organ donors and gives all of them the opportunity to opt out of the programme by completing a simple form. Research suggests that only few of the citizens will opt out. Such a programme is expected to be much more effective than a programme that simply asks people to join the donor list.

8.4.2 Framing

Prospect theory, for which Daniel Kahneman received the Nobel Prize, indicates that, from a human thinking point of view, a loss is more significant than an equivalent gain. Meaning to say that the pain of losing 1000 Dollars is bigger than the joy of gaining the same amount. We have the tendency to be risk averse when faced with a positive frame, while seeking risks when a negative frame is presented. So if we want people to choose a certain alternative, it is more effective to present the option as helping them to avoid a loss rather than to achieve gain.

Speaking to two groups of healthcare professionals and policy makers, Daniel Kahneman presented them with the following scenarios:

Group 1: Positive Framing

You are faced with a situation where 600 patients are infected by a deadly virus. There are two treatments available, A and B. Which one would you choose?

A. A treatment that will save 200 lives.
B. A treatment with 33% chance of saving all 600 people, 66% possibility of saving none.

Group 2: Negative Framing

You are faced with a situation where 600 patients are infected by a deadly virus. There are two treatments available A and B. Which one would you choose?

A. A treatment in which 400 people will die.
B. A treatment with 33% chance that no one will die, 66% probability that all 600 will die.

Treatments A and B presented to both groups are essentially identical. However treatment A was chosen by 72% of the respondent of group 1 (when it was positively framed in terms of saving 200 lives), compared to only 22% of the respondents for group 2 when it was negatively framed in terms of the 400 people who will die.

This knowledge is useful when designing policies and systems when we would like a certain behaviour to occur. If you wish to encourage the students to submit their work on time, which technique will be more effective in convincing more students to do so, giving those who hand in their work in time a bonus of 5 extra marks or imposing a penalty of deducting 5 marks for late submission? Test the validity of your answer by asking a group of students how they would respond.

8.4.3   Anchoring

This cognitive bias refers to the effect of initial, sometimes arbitrary, information on decision-making. In an experiment that was repeated around the world, the audience were shown an arbitrary number, say either 20 or 180, and asked a question "how many countries are there in the world?" Those who were shown the smaller number, 20, gave on average a smaller answer compared to those who were shown the number 180. The initial number seems to have an impact on how we judge future quantities and hence it is called the "anchor."

This has implications in a variety of scenarios. For example, the "original" price of an item on sale is a very important anchor to make the sale price look more attractive.

Chapter 9
Communication and Teamwork

"Talent wins games, but teamwork and intelligence win championships."
Michael Jordan

"A man's character may be learned from the adjectives which he habitually uses in conversation."
Mark Twain

In order to successfully and effectively Conceive, Design, Implement and Operate products, systems, and solutions, engineers often need to work in multidisciplinary teams. Effective communication skills and teamwork are essential for success in work and in life at large. This chapter will outline a number of communication and teamwork skills, techniques and strategies.

9.1   Communication Strategies

Every day, from the moment we wake up to the moment we go to sleep, we are continuously communicating. Even if we do not notice, the advertisement on billboards as we go to work or school and the websites that we visit, are always trying to communicate their messages to us. Companies want us to buy their products and they try to communicate why we need these products and how they are superior to others. We communicate with our team members, friends, parents, and children seeking their cooperation and understanding. One way or another, the purpose of communication is to communicate messages and change behaviour. Salespeople want to communicate how superior their products are (message) and want their customers to buy those products (behaviour); teachers want to communicate knowledge (message) and want their students to learn (behaviour); and politicians want to communicate their political views and policies (message) and want the citizens to vote for them (behaviour). In this context, it is fair to say that successful communication is when the right message is communicated to the right audience and it stays in their mind to create the desired behaviour.

Communication can be verbal or non-verbal. In the book titled 'The Tipping Point', Malcolm Gladwell explored a number of ideas that "stuck" and became phenomena that were talked about, shared, and eventually changed history. Inspired by Gladwell's book, Chip and Dan Heath wrote their book 'Made to Stick' and identified the features that an idea or a communicated message has in order for it to stick and result in a change of behaviour of the receiver. 'Made to Stick' utilises a framework that uses the acronym SUCCES, which stands for Simple, Unexpected, Concrete, Credible, Emotional and Stories. I added another S, Simulation and I shall describe the SUCCESS framework in this section.

Simple

We will always remember a simple message better than a complex one. The interesting thing is that, often simple is more difficult to achieve than complex. To simplify anything, one must be very well versed with all of its aspects. Apple products, for example, are known to be simple and easy to use; this simplicity is the outcome of the labour of many designers and engineers who worked tirelessly to fully comprehend how will the products to be used, and designed them for a memorable user experience. In short, simple is not easy. One of Picasso's paintings, the Bull, illustrates this concept well. Picasso drew 11 paintings of a bull starting with a life-like drawing of the bull and progressively removing parts of it in a quest to reach the essence of the bull. The eleventh painting uses only few lines to represent the bull. It is simple but Picasso reached this simplicity through his mastery of the complex.

If you have an idea, a message or a concept that you wish to communicate, you will need first to really comprehend its essence. This essence can be the basis of your communication strategy. You will need to be able to explain your concept with as few words as possible, stripping any part that is not core for the message. Simple messages are compact and profound while conveying the core essence of what needs to be communicated.

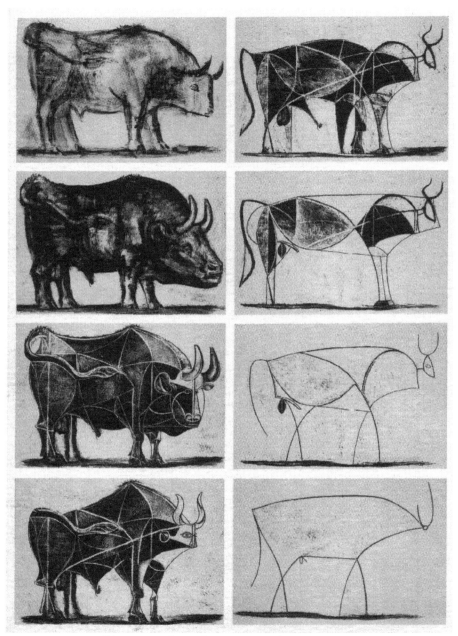

Picasso's "The Bull". (Source: takeovertime.co)

Unexpected

We explored the role the old brain plays in responding to what is perceived as a threat. This part of the brain is wired to make us quickly react to situations that are out of the ordinary, and it is responsible for the fight or flight reaction. That is why shocking, surprising, and unexpected events or information seize our attention more than expected ones. If you go for a lecture on your first day of university and your Thermodynamics lecturer arrives in blue suit and a black tie, this will not be remarkable, but if he instead arrives in a clown outfit, I bet you will never forget this event even after a long time! Always look for an unexpected angle to pitch or deliver your message. The unexpectedness can be achieved through how the message is crafted, the medium used to deliver it, or who delivers it.

Blendtec is a company that makes blenders, and in order to advertise how strong their blenders are, they made a series of videos showing their blenders in action. They did not demonstrate their blenders mixing fruits and vegetables; instead they did something highly unexpected. They filmed their blenders mixing everything from golf balls to iPhones. These videos are called "Will it blend?" I watched the "blending" of an iPhone 5S and 4 iPhone 5Cs (the full colour range) and the blender turned them into dust! This video has over 2 million views on YouTube. Will it Blend videos are also available at *http://www.willitblend.com*. This sure had an impact on me as I am planning to buy one of these blenders.

When my school wanted to put a newspaper advertisement to promote engineering as a career option, we looked for an unexpected angle to pitch that. While a typical advertisement promoting engineering may talk about the role of engineers in building different technological gadgets, we used the fact that 33% of CEOs of the world's top 500 companies have an engineering degree as the main point of the ad. According to the Business Insider, only 11% of these CEOs have a first degree in Business Administration. Many people find this to be unexpected and rather amazing. When the advertisement appeared in a local paper it had a very good impact on our intended audience and many parents and students had a reinforced view that engineering is a flexible and interesting career pathway.

Credible

In order for the message to be effectively communicated and achieve its intended impact, it needs to be credible. This means that the new brain needs to believe it. If Blendtec wants to prove that their blenders are strong, running an advertisement of them blending iPhones is not only remarkably unexpected but also credible. There are a number of ways to lend credibility to a given message. In our engineering advertisement mentioned earlier, it was necessary that we referenced the Business Insider newspaper as the party that performed the research which showed that 33% of the CEOs of world's top 500 companies are engineers. This was an independent study that the public has access to and can check out for themselves. Endorsements by users or professionals can lend credibility to some products. Toothpaste companies for example, enlist the help of dentists to endorse their products.

Barry Marshal is an Australian scientist who had a revolutionary idea. While the medical community believed that stomach ulcers are caused by stress and a spicy food diet with basically no cure, Dr Marshal had a different idea. He believed that the ulcer is caused by bacteria and are treatable by antibiotics. The prevailing wisdom was that the stomach was too acidic for any bacteria to live in it. Faced with a community that did not believe him, he did the ultimate, he infected himself with the bacteria, resulting in an ulcer in his stomach! After the disease was diagnosed, Dr Marshal treated his ulcer with antibiotics. This discovery had a huge impact on hundreds of thousands of people who have stomach ulcers and eventually earned Barry Marshal the Nobel Prize.

Concrete

It is easier to comprehend concrete concepts as compared abstract ones. This is particularly true when dealing with technological terminology. For example, if you are contemplating buying an iPhone, you will know that the 32GB option will have double the capacity of the 16GB. But do you really know how much is 16GB? Imagine communicating the capacity of the phone by how many photos or songs it can store or how many hours of video it can accommodate. Everyone can relate to the photos, songs, and videos because they represent concrete concepts.

Successful marketing campaigns and communication strategies need to differentiate their claims from those of the competitors in a concrete manner in order for them to be effective. So if the product or service that you are promoting is better, bigger, cheaper, faster or more effective than the competition, your message needs to deliver this in a clear and comprehensible manner. For example, if you want people to donate to charity and your message is intended to say that even a little bit can help vaccinate children in poor countries, the message can be "for the price of a cup of coffee, we can vaccinate 1 child against polio." Comparing the cost of polio vaccine to the price of something ubiquitous in our everyday life, such as coffee, makes the impact of a donation very clear and can lead to more people willing to donate. If the message is delivered through a visual medium, a picture of a cup of coffee next to a healthy vaccinated child can be very helpful as well.

Emotional

Earlier in the book we mentioned that emotions are processed in the middle brain. The relationship between emotions and decision-making was also examined. With this in mind, it is clear that messages that have a healthy dose of emotions in them will have a better chance of being remembered and changing one's behaviour. Research has shown that when we receive a message, we are most likely to remember and respond if this message was delivered emotionally. That is why organisations that are fighting a certain disease among children, put pictures of children on their marketing materials when they ask you for donations.

The trick is, when you develop a communication strategy to sell a product, service or idea, you need to communicate how the audience will feel when they buy your product, use your service or adopt your idea. So if the product is a phone, do not only mention the features, instead focus also on what the user will do with the phone, this would include pictures showing the users connecting with their loved ones, reaching their destinations safely (GPS feature) and having fun with friends (entertainment apps).

Stories

We all enjoy a good story. We liked it when we were kids and we continue to like it as adults. A good story represents a better way to remember and it is always nice to share. If you are running a tuition centre, a success story about an average student who managed to join an engineering course after taking math and physics classes at your centre is a good way to say that you provide quality service. The message can be even more credible if it is told by the student herself.

When telling a story, or delivering a presentation, you need to pay special attention to how you start and how you end your presentation or your story. Our brains seem to pay special attention to the beginnings and ends of stories and presentations. This is probably the reason why most of the stories that our parents told us as children began with "Once upon a time...." and ended with "....... and they live happily ever after." If you try to recall the last movie you watched, there is a good chance that you remember its beginning and end in more detail in comparison to the rest of the movie. Hollywood understands that and directors pay special attention to making the beginnings and ends of their movies more sensational.

Next time you are delivering a presentation, promoting an idea, a product or a service, you may want to begin your "show" with outlining the pain your audience is experiencing (and hopefully your product, service or idea can help alleviate it). You can use combinations of simple, unexpected, credible, concrete and/or emotional facts, data, and pictures to drive home the message. This beginning has to be short and direct so that it stays with those listening to you. The body of your presentation can discuss different features, aspects and capabilities of your idea, product or service. In concluding your presentation, you may again use a combination of simple, unexpected, credible, concrete and/or emotional messages outlining how your audience would feel after adopting your idea, product or service. This way, you will make sure that your core messages are fresh and present when your customers make their decision.

Simulation

Simulation is a powerful tool to put the audience of a communicated message in a receptive frame of mind. Whenever possible, it is useful to get the intended users to try the product, service, or idea being promoted. Producers give free samples of their products to allow consumers to experience them. Subscription-based services such as magazines or satellite channels can give free subscription for a period of time to allow the customers to experience the service. Nowadays, most of the apps have free versions that allow the users to try the app before committing to purchase the full version. The Apple

Store's Communication for Success is a very good example for this concept, where customers are encouraged to play around with the products.

Armed with the above SUCCESS framework you will be able to structure your communication strategies to achieve the desired objectives. The framework will definitely be useful if you are giving a presentation, designing a poster or billboard advertisement, directing a documentary or producing a product catalogue. Because oral presentation (accompanied by PowerPoint or Keynote slides) is one of the most common forms of communication nowadays, we shall discuss effective presentation skills here.

While you are structuring your presentation, you need to remember the SUCCESS framework and apply it. To do this, the following steps will be useful:

1. Clearly identify the objective of your presentation (message) and ask yourself what you wish the audience to do after listening to you (behaviour). If you are seeking approval for a project for example, how will your presentation provide the decision maker with the information necessary to grant the approval?

2. Understand your message fully and use this understanding to distil the essence of the message and remove any non-core parts of the message. This will make it Simple.

3. Your slides need to be simple as well. To make the slides simple, keep the slides' background white or off-white, use as few words as you can and use pictures whenever possible.

4. If appropriate, use some relevant yet Unexpected facts in your presentation. Messages like "33% of the CEOs of world's top 500 companies are engineers" fits the bill very well.

5. Make sure that the message you are presenting is both Credible and Concrete. You can do that by referencing some research done either by you or by someone who is considered an authority in the field.

6. To speak to the Emotions of your audience, tell them how good approving your proposal or purchasing your product will make them feel. Remember, it is about them not about you.

7. A good story will always help.

8. If you can get the audience to Simulate the experience of using your product or service, just do it! This can be done if you bring along a sample of your product or your previous work.

9. Remember to establish eye contact and to speak clearly. Never read from your slides.

9.2   Process Documentation

It is necessary to document the progress of the CDIO process, including and especially the failures. This is necessary to ensure that learning is taking place, as well as to enable continual quality improvement. The documentation represents a legal necessity as well. Good documentation can play a key role in establishing intellectual property in the case of a patent dispute. In this section, we shall explore three main documentation methods, namely the logbook, minutes of meetings, and technical reports; along with tips on e-mails and the use of operation manuals.

9.2.1   Logbooks

A logbook is a book that a team member keeps to document the progress of the project being worked on. People keep logbooks to document the progress of their postgraduate projects, technical projects and business projects. For the logbook to preserve its legal relevance in case of future dispute, the following are necessary:

1. The logbook should consist of pre-numbered pages that are not easily torn off.

2. All entries should be dated and signed.

3. Only permanent pen is to be used to record the entries.

4. No correction ink is to be used. If corrections need to be done, the entry can be struck off.

5. No empty pages should be left in the logbook. Any empty page should be struck off.

9.2.2   Minutes of Meetings

Meetings are important team activities especially when decisions are to be made, tasks are distributed and progress is reported. Thus it is very necessary to run meetings and document them in a structured and effective manner. It is highly recommended that a team member be designated as the chairperson of the meeting. This is normally the project manager or the team leader. The meeting should have a designated scribe or secretary who will take the minutes. Team members may take on this role alternatively.

The process of running a structured meeting is described below:

1. The secretary sends the meeting agenda to all the team members, preferably 2 days before the meeting. The agenda is a document that outlines the items that the meeting will discuss. The agenda is set based on what the meeting intends to achieve, team members may request certain items to be added to the agenda.

2. The first item on the agenda should be to review and approve the minutes of the previous meeting. This is to ensure that all the team members accept the minutes as a correct and accurate representation of what was agreed upon in the previous meeting.

3. The second item on a meeting agenda is the "matters arising" from the previous meeting. Matters arising are action items requested of the team members and they need to report on their progress.

4. At the time of meeting, the chairperson calls the meeting to order and follows the agenda. After approving the minutes of the previous meeting and reporting on the matters arising, the meeting then goes through the new items on the agenda.

5. The scribe records the discussion focusing on decisions made, which need to be recorded clearly, and action items as they will represent the matters arising for the next meeting.

9.2.3 Technical Reports

Technical reports are very important documents that outline the work done, and the main findings and impact of that work. It is important to avoid any plagiarism when writing a report and any work of others used should be properly cited and attributed to the source. Nowadays, there are software tools to help check the submitted work against other available and published work.

A technical report usually follows the structure below:

1. **Cover Page**
    The cover page should have a clear and concise title of the report together with the names of the authors who have substantially contributed to the work.

2. **Executive Summary or Abstract**
    This section gives a summary of the work undertaken, its objectives, methodology, and its major findings.

3. **Introduction**
    The introduction section (it can be an introduction chapter for a longer report) contains basic information about the work undertaken, including the background and objectives, stakeholders, and success measures.

4. **Literature Review**
    In this part of the report, the available information about the work is outlined and critically analysed. A good literature review surveys the available books, journals, patent databases, conference proceedings and other technical reports. The outcome of a good literature review is the establishment of the edge of the available knowledge and the identification of any knowledge gaps and challenges that are delaying progress in the field.

5. **Methodology**
    This section is dedicated to describe the methodology that is used to achieve the objectives. This could be experimental, computational, mathematical, statistical or combination of these methods. The section also describes how the data will be collected and analysed and clearly outlines the limitations of the methodology.

6. **Results and Findings**
    All major findings of the work done are documented in this section. These findings can be reported in the form of graphs, tables and figures for easy analysis.

7. **Discussion**

In this section, the results are critically discussed and compared with other results available in the literature. The impact of the results of the work is elucidated and suggestions for future work may be recommended.

8. **Appendices**

This is an optional section where detailed derivations of some equations used in the report, and other materials relevant to the report but are too lengthy to be part of the main report body can be included.

9. **References**

All the references used in the report need to be listed here. There are numbers of systems to do the references, including Harvard, APA and IEEE. Referencing system should be consistent throughout the report.

9.2.4 Communication via E-mail

E-mails are a very important form of communication. In this section, some guidelines for the effective use of emails are outlined

1. Use an e-mail ID that represents you, your team or your company. E-mail IDs such as *superblur@hotmail.com*, may be cool but they will not help you create the impression which you want to give. This is especially true if you e-mail bankers, funders, academic institutions and government agencies.

2. When sending an e-mail put the intended recipients in the "to" field. Those who are carbon-copied for their information should have their e-mails in the "cc" field. "Bcc" or blind carbon-copy can be used to e-mail a large number of recipients while keeping a personalised feel. You will also avoid the recipients replying to all and creating nuisance for others who would then receive unwanted e-mails.

3. Choose a concise and representative "subject" for your e-mail.

4. Avoid putting images in the e-mail as most e-mail providers will filter them out.

5. Avoid sending large attachments unless necessary. Nobody enjoys having their inbox flooded by large messages.

6. Avoid capitalising words and paragraphs in your e-mail. Capitalised words are difficult to read and they give the recipients an impression that you are shouting at them.

7. Although e-mails are a quick and efficient way of communication, they represent a poor way of communicating when emotions are high. When you are angry, do not reply e-mails. You may just make things worse. You can reply the following day when you have cooled down or you may even pick up the phone and speak to the person who e-mailed you to explain your stance.

8. Use a neat signature that contains your name, position and contact details.

9.2.5   Operation Manuals

Operation Manuals are very interesting documents that users are supposed to read before using their new products. In the human-centred design chapter, we postulated that a well-designed product should be used with minimum reference to instructions. However, many products will still need operation manuals, these can range from an IKEA table that needs to be assembled by the customer to an airplane with an operation manual that describes various aspects of its operation and maintenance.

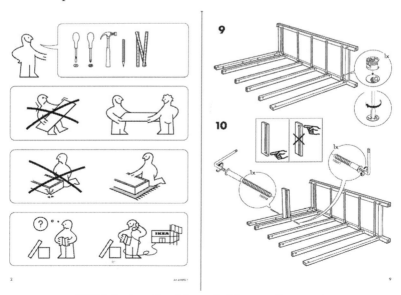

IKEA Instruction Manuals (Source: ikea.com)

Operation Manuals may serve legal needs as well; it is where the manufacturers describe the way a product should (or should not) be used. In any case, operation manuals should be developed with the user in mind. This means that they should be clear, concise, and straight to the point. My favourite example is the IKEA manuals; they contain very few words, and anyone without technical expertise can use them to assemble the furniture purchased.

9.3   Teamwork

Teamwork is a necessary collaborative skill; it is very difficult to imagine success without it. The challenges that are presented by today's complex environment can only be addressed by highly effective teams of people from different walks of life. Although teamwork is one of the highly sought-after skills, employers around the world are complaining that academic programmes are unable to equip the graduates with the necessary skills to work well in teams. This section attempts to provide a simple and clear methodology to develop skills and techniques that can go a long way in ensuring sustainable team success.

9.3.1   Selecting the Team Members

Needing a team to complete a given task is an acknowledgement that, no matter how clever an individual is, she or he will most likely not have all the necessary skills and resources to complete the task. Successful teams comprise of motivated people who have the necessary skills and attitudes to address the challenge at hand. Team members complement each other and bring to the table the diversity that is necessary to encourage creativity and amplify the team's effectiveness. Successful teams have members who possess diversified skills that can be described in the cognitive, psychomotor, and affective domains, as well as other relevant attributes for the specific function or role that they are playing. The cognitive, psychomotor, and affective domains are commonly referred to as the skills of the head, hand, and heart, respectively.

The cognitive domain is associated with the knowledge required to fulfil a specific function within the team. An example of this would be knowledge in international law, mathematics, or biology. The psychomotor domain involves the specific skills required to perform a given function such as operating a certain machine or using specific software. The affective domain includes attitudes and emotions the person has to have in order to effectively work in the team. This may include self-awareness and empathy. For example, a certain team member may be required to have knowledge in international business law (cognitive), skills to operate the company proprietary software (psychomotor), as well as good communication skills and empathy (affective). Other attributes may include the person's height, weight, or professional qualifications. The various functions of the team members are dependent upon the type of project.

Take the Taylor's Racing Team for example, in the team there are specific functions, such as the driver, team manager, and mechanic, and each has an important role to play in the team's success. These roles require specific cognitive, psychomotor and affective attributes along with other relevant attributes. These are summarised in the following table.

Team Members' Attributes

Function	Knowledge Cognitive	Skills Psychomotor	Attitudes/ Feelings Affective	Attribute
Driver		Driving		Lightweight Race License
Team Manager	Race Regulations		Empathy	
Mechanic	Engine Principles	Engine Repair		

The collection of knowledge, skills and attitudes of team members, if integrated in a synergic manner, result in the development of the core competencies of a team. A core competency of a team or a company is a distinct ability that a team has that contributes to its success and results in integrating and amplifying the strengths of the individual team members.

9.3.2   Core Competency

To define core competency, let us first define both words individually. According to the Oxford Learners Dictionary, core is defined as the most important or central part of something; competency is defined as a skill that is required in a particular job or for a particular task. So, in general, a core competency is an ability or skill that is central to the identity of the team and is a result of the combined skills, techniques, and culture of the team, thus giving the team a competitive advantage.

For example, Apple Inc.'s core competency is its attractive design and the ease of use of its products. While the core competency of Dell can be considered to be its efficient and integrated production system.

9.3.3 Organisation Chart

When working in big teams or when engaged in official work that requires clear documentation of individual roles, an organisation chart is necessary to display the structure of the project team, team members, their functions, as well as their relationship and the lines of authority within the team. The organisation chart can be used to streamline and focus the core competencies of the team, leading toward sustainable delivery of value. It can also be used to simplify the decision making process.

The most common organisational charts used in organisations, projects or businesses are the hierarchical type and the matrix type. The hierarchical type of organisation chart has a top down order of organising the team members and represents a more traditional way of running a team. The team leader would be at the top of the organisation chart, a few people would report to her (him) and there would be a few others reporting to them. The hierarchical type organisation chart can be detailed and complicated with many layers depending on the organisational needs.

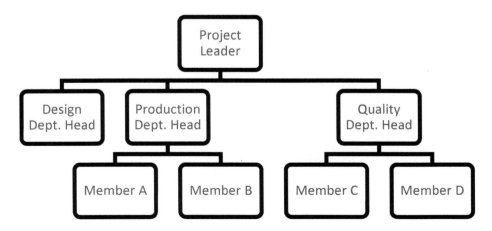

Hierarchy organisation chart

The matrix type organisation chart reflects a more flexible approach to resources allocation. It has a leader at the top, managers for various functions and managers for different projects. Managers of different functions such as design, planning, and production are expected to allocate resources and support different projects.

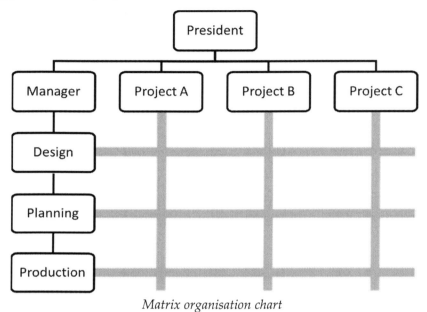

Matrix organisation chart

9.3.4 Team Evolution

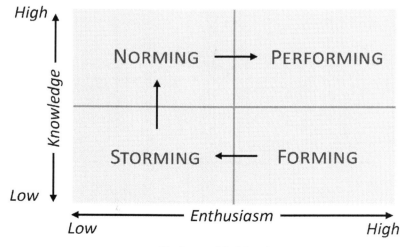

Tuckman Model

Teamwork is a complex human interaction that requires certain skills that can be learned and developed over time. This starts with understanding the various stages a group of people who are working together as a team normally go through. The Tuckman Model is one of the best models that describes how a team evolves and can provide a useful insight into human interaction in a team.

1. Forming

Forming is the first stage of team evolution. The team members are normally selected because of their capabilities and skills related to the project. During this stage, the enthusiasm among the team members is high, but the knowledge of the tasks assigned to them and their teammates, as well as the team dynamics, is usually low.

2. Storming

As time goes by and tasks accumulate, enthusiasm wears off but knowledge may not necessarily grow. The combination of low enthusiasm and low knowledge leads to stress and conflict, and different team members respond differently; some will be disappointed and confused, whereas some will just withdraw. The storming stage is a difficult one but it is necessary in order to move into the remaining team evolution stages. The key is to be able to recognise the storming stage when it happens and go through it swiftly.

3. Norming

The norming stage is where leadership potential shines. Normally what happens is that some (at least one) team members take charge and pulls the team members towards achieving the team's objectives. In this stage, enthusiasm may still be low, but knowledge and respect are definitely on the rise. Team members will start to taste success as they meet task deadlines. This stage is necessary to lay down the foundation for the next stage.

4. Performing

Building on the momentum of the norming stage, the performing stage features high knowledge and high enthusiasm. Arriving at this stage is a hallmark of successful teamwork. In this stage, respect and appreciation for team members' individual qualities are prevalent. Team members also have an accurate self-assessment of their own strengths and capabilities.

The table below shows how a typical student team evolves over a period of a 14-week semester at Taylor's University. The students work in multidisciplinary teams to Conceive, Design, Implement and Operate an artefact which can be a racing car, an ergonomic chair, a solar boat, a robot or a vending machine.

Week	Student	Team Stage	Team Activity	Project Stage
1	Overwhelmed	-	-	-
2	Finding the way	Forming	Casual meeting	-
3	Knowing the team	Forming	Casual meeting	Sketch (C)
4	Disappointed/ Misunderstood	Forming	Serious meeting	Sketches (C)
5	Blaming/ blamed	Storming	Serious meeting	Initial Design (D)
6	Complaining/ withdrawing	Storming	Conflict	Improved Design (D)
7	Forced to work it out	Storming	Conflict	Improved Design (D)
8	Leading/ following/ kicked out	Storming	Leadership emerging	Improved Design (D)
9	Leading/ following	Norming	Negotiating	Final Design (D)
10	Participating	Norming	Negotiating	Building (I)
11	Actively participating	Norming	Role acceptance	Building (I)
12	Understanding	Performing	Role understanding	Building (I)
13	Respecting	Performing	Mutual Respect	Building (I)
14	Appreciating	Performing	Appreciation	Completing (O)

9.3.5 Team Building Exercises

After selecting the team members, we can accelerate and boost the team performance through variety of team building exercises. These are activities that the team members can do together. Normally, team members will do some fun exercises as a group. These can include solving some interesting mental or physical challenges or even cooking. Through these exercises, team members get to appreciate each other and the diversity that they bring to the team. Team building can be done with newly formed teams or pre-existing teams that are given new tasks or wish to rekindle the team spirit.

As mentioned in the emotional intelligence chapter, our school team went for the "Shoot the Boss!" activity in 2013 before accreditation. This activity also served as a team-building exercise. The accreditation is an extremely important event that ensures that our graduates will obtain internationally recognised degrees. To achieve better team spirit among our academic team, we designed a fun activity where we went to a paintball court and each member of the staff was asked 3 in-depth questions about the accreditation to gauge their knowledge and their level involvement in the accreditation process. If the staff member fails to answer at least 2 out of 3 questions correctly, the staff needs to take the marker and shoot me (the dean of the school). Because most of the staff did not want to shoot me (or so I hoped), they really prepared for this exercise. Still some of them did not manage to answer 2 out of 3 questions right and they had to play the role of the firing squad. While most of them missed me on purpose, the exercise really brought the team closer together. This exercise enabled us to identify a few areas for improvement, created alignment among the staff, and most importantly communicated the commitment of the school's management in a very clear way. Since then we have successfully achieved our accreditation. To our delight, the accreditation report identified a number of strengths for our school, among those was "highly motivated staff and students". Now, other departments in the university are asking us to facilitate similar team building sessions for their staff.

9.4 Connections and Networking

The ability to establish diverse connections and networking is a key skill for success. Networks can be real or virtual, personal or professional and they all need to be purposefully built and nurtured to be able to add value to the individuals within them. Virtual networks such as Facebook, LinkedIn and Research Gate are growing in popularity, and employers nowadays check the Facebook profiles of their future employees to ensure that these job applicants represent a cultural fit for their organisations. It is very important to keep a decent profile on networks such as Facebook. LinkedIn provides a very professional platform where individuals can keep their employment history and acquired skills up to date. A growing number of human resource managers use LinkedIn as a recruitment platform when they are looking for suitable job candidates.

Networking can happen the good old way as well - in person or face-to-face networking. Professional bodies regulating the work of different professions, as well as conferences and trade shows, are good platforms to meet people with similar interests. Make sure that you carry a simple and nicely designed name card with fonts that are easily read.

Name Card

1. An effective communication strategy is described by the **SUCCESS** framework. Messages that reach their intended audience and achieve the desired change of behaviour, are often found to be:

   Simple

   Unexpected

   Credible

   Concrete

   Emotional

   Told in a Story

   Simulated

2. Team evolution is represented by the following stages:

   Forming

   Storming

   Norming

   Performing

Chapter 9
Communication and Teamwork

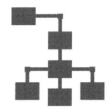

Chapter 10
Managing Projects for Success

"The most efficient way to produce anything is to bring together under one management as many as possible of the activities needed to turn out the product."

Peter Drucker

Projects are important human activities. They represent non-repetitive and dynamic endeavours that involve the orchestration of a variety of resources, including human, financial, technical and legal resources; to achieve specific objectives within a certain timeframe and budget. Each project is different and brings its own challenges and requirements; hence project management is both a science and an art. A project can be organising a party or a family holiday; renovating a house; or constructing a nuclear reactor. Managing projects is an integral part of the CDIO process and it plays a key role in bringing to life complex ideas that can range from developing a factory to manufacturing thousands of bicycles, or deploying a once-off space shuttle, to working to increase sales or improve customer satisfaction. Hence, managing successful projects can be defined as achieving the project objectives in time and on budget. This chapter will explore the basic project management tools and present them in a simplified and practical manner.

The typical project management process or project life cycle goes through the following phases:

1. Initiation
2. Planning
3. Execution

4. Monitoring and Controlling
5. Closing

These phases are described in the sections below.

10.1   Project Initiating

Project initiation is the first stage, and it takes place before the project is launched. During this stage, priority analysis, feasibility studies, and complexity analysis are performed. If a project is deemed feasible, the project scope, description and objectives are finalised and documented. This is achieved through clearly understanding why the project is undertaken and how success is measured. In this stage the following questions can be very useful:

1. **Why is the project being undertaken? What is the gap that the project is closing or what is the opportunity that the project is realising?**

    An example for a gap can be "Customer satisfaction is down by 20%," while an example of an opportunity may be "There is a demand for a smart phone application that can track the level of air pollution."

2. **How long has this gap existed, or how long is the opportunity expected to last?**

3. **What would be the impact of not addressing the gap or harnessing the opportunity?**

    The impact could be a further decrease in customer satisfaction and/or loss of market share.

4. **What causes the gap or the opportunity?**

    For example, the decline in customer satisfaction can be due to the poor quality of the company's products or the lack of training of the customer service staff.

5. **What are the project deliverables and objectives and what is the project scope?**

    The project Objectives should be S.M.A.R.T. They should be Specific, Measurable, Attainable, Realistic and Time-bound. The Scope of the project refers to the extent of the project. For example, a project to address the gap in point 1 above can have the objective of improving satisfaction level of online customers (Specific customer segment) by 20% (Measurable) within 3 months (Time-bound). The Objective is Attainable and Realistic as well. The Scope of the project refers to the extent to which the project will go to achieve the objective(s). This can be, for example, "The project will focus on training the customer service staff" or "Implement a Six Sigma quality assurance programme to improve the quality of the products."

6. **What are the resources the project needs to succeed?**

7. **Who are the Stakeholders of the project?**
    Stakeholders of a project are individuals, groups or entities that are affected or impacted by the project, and can affect or impact its successful completion. One of the key tasks and responsibilities of the project team is to identify, classify and engage the project stakeholders. If the project, for example, involves establishing a new branch of a bank, the stakeholders can include the customers, staff, the central bank which needs to issues a permit, businesses and residents surrounding the new branch, suppliers of banking equipment and ATM machines, as well as the competing banks. Every one of these stakeholders may affect the project and be affected by it negatively, positively or both. For example, residents living close to the new bank branch may benefit from the conveniences that bank brings along but they may suffer from traffic congestion as more customers start to patronise the new bank. Likewise, these residents may opt to lobby the authorities against issuing the permit to the bank to operate the new branch. It is clear that the project team needs to consider matters and relationships beyond the obvious of kitting the building and staffing the branch. Project management and stakeholders' management require a systems thinking approach that integrates the various aspects of the project to ensure that the project will be a sustainable success.

8. **What is the level of complexity of the project?**

9. **What are the risks associated with this project?**

10. **What are the policies and procedures that can affect the project?**

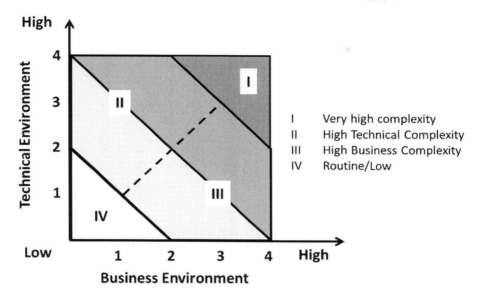

Complexity Assessment Graph

Answering the above 10 questions can help guide the thinking process and help prepare the project proposal. The initiation phase should result in the identification of clear objectives and deliverables and how to go about achieving them. This should be agreed upon and signed off by the project sponsor or customer. A project's sponsor is the individual that controls schedule, scope, budget and quality of project; while the customer represent the individual or group that either ordered the project or receives the first line of benefit from the project or both.

10.2 Project Planning

Once the project is approved, the project manager and the core team will start working on more detailed planning. This includes identifying and recruiting other project team members, and estimating the total project cost and duration. The planning stage features the use of a number of tools to estimate cost and duration for the project. These include the project's budget, Gantt chart, Work Breakdown Structure and Network Diagrams. Project planning also involves the management and mitigation of any risks that may negatively impact the project.

It is useful here to define some key project elements that are used by project management professionals. These terms are listed below:

Task: A distinct piece of work that represents part of the project and is assigned to an individual or a group. A project can be broken down to specific tasks and the timely and successful completion of different tasks is necessary for other project-related tasks to be completed.

Deliverable: A clear and measurable outcome of a successful task. For example, a task can be "To install 3 ATM machines in the new branch." A deliverable will be "3 ATM machines are operational by the 3rd of January at noon."

Lag: A time delay between subsequent tasks. For example, if the first task is to plaster a wall and the second task is to paint that same wall, a time Lag to allow the wall to thoroughly dry is necessary.

Milestone: A major and visible progress achieved on the project schedule. If you are building a house, milestones can include completing the foundation, completing the structure, exterior finishing and interior design.

10.2.1 Cost and Resources Estimation

The project cost is a key planning component. It represents an account of all the costs that the project will require in order to be successfully completed. This includes the cost of human resources and other direct and indirect costs. This has to be aligned with the budget available for the customer, and will drive a variety of decisions such as the choice of materials and other implementation techniques.

10.2.2 Time Estimation

Gantt Chart

Gantt charts are among the most essential project management tools. A Gantt chart represents the schedule of different project tasks and deliverables as a function of time. If kept updated, it can give an overall knowledge of the progress of the project.

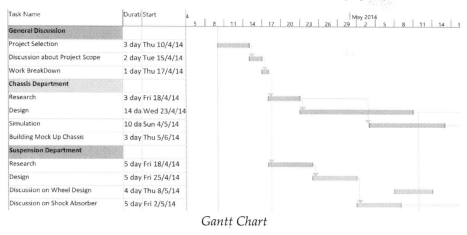

Task Name	Durati	Start
General Discussion		
Project Selection	3 day	Thu 10/4/14
Discussion about Project Scope	2 day	Tue 15/4/14
Work BreakDown	1 day	Thu 17/4/14
Chassis Department		
Research	3 day	Fri 18/4/14
Design	14 da	Wed 23/4/14
Simulation	10 da	Sun 4/5/14
Building Mock Up Chassis	3 day	Thu 5/6/14
Suspension Department		
Research	5 day	Fri 18/4/14
Design	5 day	Fri 25/4/14
Discussion on Wheel Design	4 day	Thu 8/5/14
Discussion on Shock Absorber	5 day	Fri 2/5/14

Gantt Chart

Work Breakdown Structure

This is a structural task and deliverable way of describing a project. It is a useful project management and systems thinking tool. It helps decompose a system or projects based on the flow of materials and products, data, and information, or services. An example of WBS is given below.

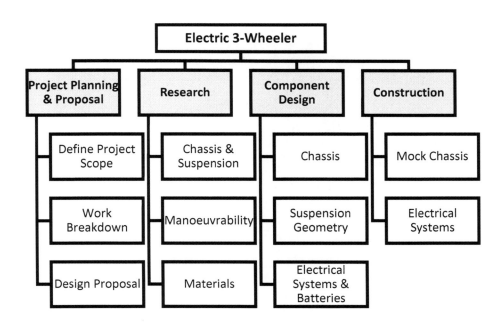

An Example of a Work Breakdown Structure

To develop useful and easy-to-use WBS, the following design features need to be observed:

1. Deliverables Driven

The WBS should focus on planning deliverables and outcomes rather than just tasks and actions.

2. 100% rule

This rule indicates that the WBS should encompass 100% of the work and deliverables defined by the project scope. This assumes that proper work has been done to ensure no major component is missed. One way to ensure that is to check the projects from the materials and products flow point of view, information and data point of view, as well as services point of view.

3. Mutual Exclusivity

There should be no overlap in scope definition between different elements of a work breakdown structure. Whenever overlap is experienced, the scope should be broken down.

4. WBS Elements' Coding

Elements of the WBS need to be coded in a meaningful way and the level of details can be fine-tuned so that the terminal elements (elements that are not subdivided) are not too big or too small. An example of the coding is given below.

1. Product Manufacturing
1.1   Customer Specification
1.2   Parts Manufacturing
1.3   Parts Assembly
1.4   Quality Control
2. Marketing
2.1   Media
2.2   Marketing Strategy
2.3   Control & Assessment
3. Finance
3.1   Budget
3.2   Source for Material Suppliers

Network Diagram

This diagram depicts the logical sequence and concurrence of various deliverables, tasks, subtasks and milestones in a project by showing different tasks in relation to each other. A very simple example is shown below.

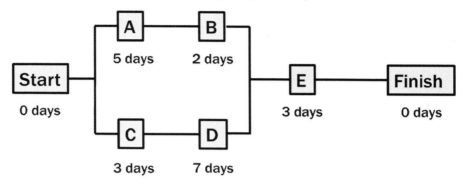

Network Diagram

The flow through which a series of activities that depend on one another and are connected in sequence is called a Path. The path that takes the longest time in the network diagram is referred to as the Critical Path. It provides an indication of how long it would take to complete the project. Any delay along the critical path will affect the entire project!

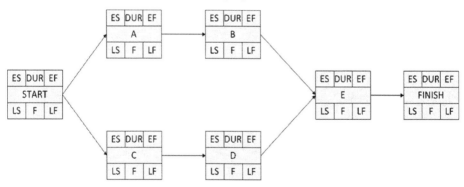

ES   : Earliest Start
DUR : Duration
EF   : Earliest Finish
LS   : Latest Start
F   : Float
LF   : Latest Finish

In the network diagram shown, task A and C can start at the same time. Task B can start only after task A is completed, that is why A is predecessor for B and B is successor for A. Task E can start after both tasks B and D have been completed, it has two predecessors. The method to identify the time required by an activity is called Activity on the Node (AoN). This gives a quick estimation of the critical path of a project. In the case of the network diagram above, the critical path is CDE (13 days). A more detailed way to estimate a project timeline is the Precedence Diagramming Method (PDM) which is demonstrated on the network diagram. Besides identifying the Critical Path, this method can help identify dependencies, lead and lag times, as well as slack or float.

Float (or slack) is the amount of time an activity can be delayed without causing a delay in the subsequent activity or the project completion date. There are two types of float, free float, which is the amount of time an activity can be delayed without impacting the ES of a successor activity, and total float, which is the total amount of time an activity can be delayed without impacting the project completing date. Generally, the float can be defined by the equation below.

$$F = LS - ES$$

Time estimates based on dependency can be obtained via the forward pass and backward pass method. The forward pass method is used to obtain Earliest Start and Earliest Finish while the backward pass method is used to obtain Latest Start and Latest Finish.

Forward Pass

$$ES\ (Current) + DUR\ (Current) = EF\ (Successor)$$

ES   : Earliest time for an activity to start

EF   : Earliest time for an activity to finish

DUR   : The anticipated duration of the activity

EF from predecessor activity is brought on as ES for successor activity

Backward Pass

LS (Current) – DUR (Predecessor) = LS (Predecessor)

LS   : Latest time for an activity to start

LF   : Latest time for an activity to finish

Joins are the activities that are connected to more than one activity, either upstream or downstream. When performing a forward pass, use highest value of ES (Successor). For backward pass the lowest value of LS (Predecessor) should be used.

In order to shorten the critical path, project managers often resort to the measures below:

1. Fast tracking: Shortening the time taken to complete an activity by converting series activities to parallel activities.

2. Crashing: Shortening the time taken to complete activity by utilising more resources.

3. Reduce scope.

10.2.3 Risk Management

Risk is often referred to outcomes that have negative impacts on the achievement of the project's objectives. When dealing with risks, one should consider the probability an event occurring, its impact, the range of possible outcomes, as well as expected timing and frequency. Past lessons learnt can inform the risk management plan. Some projects have a risk management committee that meets regularly to ensure alignment and standardisation in responses and communication.

The strategies to manage the risk includes the following:

1. **Risk Evasion**: Develop a conservative plan to avoid the risk altogether. This is also called risk aversion and may increase the time and cost of a project and result in missing some opportunities.

2. **Risk Mitigation**: Plan to reduce the risk through diversifying the supply chains, for example, to mitigate the risk of delay in supply of source materials.

3. **Risk Acceptance and Monitoring**: If there is no alternative, the risk is simply accepted and monitored.

4. **Risk Transfer**: Purchase insurance.

10.3   Project Executing

Armed with the project description, goals, cost, and duration, project managers can start the various tasks to execute the project. During this stage many tasks need to be performed, including staffing, purchasing and procurement, and sub-contracting.

Procurement of goods, materials and services at the right price and quality is an essential component of the success of any project. In a globally connected economy, appreciating and understanding the behaviour of supply chains will go a long way in ensuring the delivery of the project objectives. The procurement process typically involves the following steps:

1. Needs identification
2. Supplier selection
3. Negotiation
4. Logistics and quality management

10.4   Project Monitoring and Controlling

As time progresses, project managers spend a fair bit of their time monitoring the progress of various components of the project and making decisions to ensure that the project remains on track in terms of delivering the agreed upon goals on time and on budget. More often than not, change happens on the ground and the project plans need to be altered in response to those changes. Project managers need to dynamically respond to changes in the environment while maintaining communication channels with all the stakeholders and ensuring compliance with the legal, technical, environmental and financial requirements.

10.5   Project Closing

No project is successfully completed without a proper closing stage. In this stage the project is handed over to the project owner and the project team needs to ensure that the handover process is well documented and ensuring the satisfaction of various stakeholders. The closing process includes a review and assessment of the project performance and the documentation of any learning to be incorporated in future projects.

10.6   Project Stakeholders Management

Stakeholders are defined as any group, organisation or individual(s) who are affected by, or can affect, the success of your project and/or operation. Stakeholders can be internal or external, and the stakeholders' landscape can be very dynamic and a new group can join or leave the stakeholders' list.

When studying the stakeholders, it is useful to ask the following 5 questions:

1. Who are your stakeholders?
2. What do your stakeholders value?
3. Are you providing that?
4. How far are you from achieving stakeholders' satisfaction?
5. Do you have plans to close the gap?

The ICE framework for stakeholders' management involves three stages, namely to Identify, Classify and Engage with the stakeholders.

Stakeholders Identification

There are a number of ways to draw the stakeholders' map. The choice of the suitable map is dependent on the purpose of the stakeholder analysis. The map can be drawn by looking at the network diagram of a project; everyone who performs or can affect a task is a stakeholder. A more generic stakeholders' map is given below.

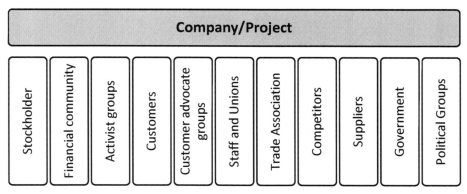

A Map of the Stakeholders of a Company/Project

Stakeholders Classification

Stakeholder identification often results in a long list of all the stakeholders that can affect, or be affected by the project. Hence, it is essential to classify the stakeholders to prioritise the engagement process. This can be done based on the interest and power of each stakeholder.

Stakeholder's interest determines if the stakeholder is more likely to cooperate or be a threat, while the stakeholder's power determines the level of the cooperation or threat presented.

Stakeholders Classification

Stakeholder's Potential for Support/Threat to an Organisation

Stakeholders can also be scored based on their power, proximity and urgency. The stakeholders can be ranked based on the total score, the higher the score the higher the importance of a stakeholder.

Power

1: Cannot cause much change.
5: Has the power to stop the project or activity.

Proximity

1: Remotely working with the project or activity.
5: Directly working with the project or activity.

Urgency

1: Little action outside the routine is required.
5: Action needs to be taken immediately.

Stakeholders Engagement

After classifying the stakeholders based on their importance and whether they are supportive, non-supportive, marginal or mixed-blessing, an engagement strategy can be drawn out. The communication is the essence of the engagement strategy and the SUCCESS framework, as discussed earlier, is a good one to adopt here.

Supportive stakeholders are those whose interests are aligned with those of the project and represent no threat. They can be advocates for the project and the engagement plan should focus on empowering them to do so. An example would be the stakeholders who have a direct interest in the success of the project.

Marginal stakeholders have little interest in the success of the project but they represent no threat either. They can be monitored for any changes in their stand. There is potential to win them over.

Mixed-blessing stakeholders are those who have high interest in collaborating with the project team and at the same time, if they are not happy or satisfied, can represent a real threat. Continuous monitoring and communication are necessary to ensure that they are positively engaged. Examples of these stakeholders are governmental agencies that have an interest in the given project and they have the power to shut it down, for example through cancellation of permits.

Non-supportive stakeholders are those who represent a threat and the potential to collaborate with them is limited. Competitors fall in this category and we need to monitor them and defend the interest of the project if necessary.

Chapter 11
Entrepreneurship and Innovation

"Innovation is the specific instrument of entrepreneurship. The act that endows resources with a new capacity to create wealth."

Peter Drucker

Entrepreneurs and innovators are everywhere. They continuously surprise us and push the limits in their pursuit of uncovering, developing, and delivering value. They succeed at times and fail at others, but they always continue to try. In 2013, I delivered a Massive Open Online Course (MOOC) on Entrepreneurship. This course was one of the first MOOCs in Asia and it was a success with more than 2,500 students from more than 120 countries. A few months after delivering the course, my son asked me, "Dad, you are not a businessman, so how come you are teaching entrepreneurship?" My answer was that I see entrepreneurship as a frame of mind. Although entrepreneurship is often associated with business activities, teachers, doctors, students, and community leaders can be entrepreneurial when they go about doing their jobs effectively and differently, adding value and reaching new levels of performance. I hope that had I managed to convince him!

Entrepreneurship and innovation skills are mental models and thinking habits that can be developed and nurtured. Wiring of these mental models and thinking habits can generate a mindset that focuses on the identification of opportunities and delivering value and is rooted in the persistence to push the limit to continually learn and grow. Although entrepreneurship is often associated with business activities and money-making, entrepreneurship at its heart is about finding one's inner calling, creating value and building a better world using innovative methods. So while founders of great businesses are successful entrepreneurs who create value and provide employment, entrepreneurship can be anything and everywhere including at not-for-profit organisations. It can be practiced at home, in schools, in government agencies, hospitals, and sport clubs. A mother who keeps her family healthy and well fed despite the difficulties in the environment can be considered an entrepreneur. A teacher who utilises innovative techniques to improve the students' learning experience can be regarded an entrepreneur too.

We know that every person is unique. This is literally true as the genetic composition of each one of us is different from that of everyone else who has ever lived on this earth. This makes everyone's potential unique and distinctive. In order to achieve individual fulfilment and happiness as well as collective diversity, this uniqueness needs to be nurtured and expressed, as this will help us find our individual purpose and inner calling, and in the process will make our society more diverse and resilient. Entrepreneurship as a mindset as well as activity provides the avenue for human uniqueness to flourish and enables individuals to realise their potential. This century can be considered to be the century of entrepreneurship and innovation, where human uniqueness is celebrated and where the challenges we are facing can only be addressed using totally different approaches from what we have been utilising thus far. This is also the age when technology empowers individuals to pursue the mastery of their destiny. Whether you aim to have your own business and make profit, or seek employment in an existing set-up, entrepreneurial thinking will be a valuable asset. As a matter of fact, employers around the world are searching for innovative and entrepreneurial individuals to fill the vacancies at their companies and organisations.

This chapter explores entrepreneurship and innovation as a mindset that seeks opportunity and value. The chapter also introduces techniques that can be used to create value and realise opportunities in a variety of contexts.

11.1 Philosophy of Entrepreneurship

Entrepreneurs are optimistic and pragmatic individuals who believe in creating a better world. Often they take risks to expand the existing paradigms and they are motivated by making a difference as a way of making money. They see challenges as opportunities and solutions in disguise.

Entrepreneurs are innovative and creative individuals that inspire change and add value in all walks of life.

In 1990, Jerry Sternin was sent by the "Save the Children" organisation to fight severe malnutrition in rural communities of Vietnam. The Vietnamese foreign minister gave him just six months to make a difference. Studying the situation, he realised that malnutrition in Vietnam is caused by poor sanitation, poverty, and lack of education. These are chronic challenges that had plagued the country for decades and if Jerry were to make any difference in six months, he needed to Conceive an innovative and entrepreneurial solution. He simply asked a question, are there any poor families whose children were bigger and healthier than the typical child? When the research team started to collect data to answer this question, they indeed found some children who, despite poverty, were healthy and well fed. Observing what the mothers of the healthier children were doing differently, the team noticed that these mothers were feeding their children smaller portions of food, more often throughout the day. Mothers of healthy children were also collecting brine shrimps from the rice paddy fields and sweet potatoes grown in their gardens and adding these to their daily soups or rice dishes (even though most people avoided these "low class" foods). When serving their children, these mothers were observed to scoop from the bottom of the pot, making sure that the children ate the shrimp and greens that had settled during cooking. Empowered by this revelation, Jerry was ready to Design a nutritional programme that was based on the successful mothers' feeding routines. The Implementation stage of this project took place as Jerry and his team prepared the communication plan to show the mothers that the solution was in their hands and they can do something to improve the health of their children. Operation was done as Jerry organised group cooking sessions to share the best practices with the mothers and ensure that they were well trained.

The project resulted in the sustainable rehabilitation of hundreds of malnourished children. The reason I described this project using the Conceive, Design, Implement, Operate (CDIO) terminology is to emphasise that CDIO represents a systems approach to entrepreneurship that can be learned, analysed and improved.

11.2   Business Value

Creating value is at the heart of innovation and entrepreneurship. Value is created through producing products, goods, and services that perform a needed task and generally make life easier and more enjoyable. Business value is the worth created by the entrepreneurial activities as perceived by different stakeholders. A sustainable business is one that is able to create value for the customers, shareholders, employees, suppliers, partners and the society at large. Understanding the value creation and management can enable the development of sustainable business models. It is important to know that different stakeholders' values are often integrated. The next sections provide descriptions of business value as perceived by different stakeholders.

11.2.1   Customer Value (Value Proposition)

Value proposition summarises the unique value(s) a business promises to its intended customers. To develop differentiated, compelling, and unique value propositions, entrepreneurs need to be very clear on the following:

1.   Who is the customer? What market segment is being addressed?

2.   What do the customers value, need, and desire?

3.   How is the entrepreneur addressing the customers' needs and wants?

4.   How much are the customers willing to pay for the product or service?

   A good value proposition has to be differentiated from those offered by the competitors and it can be based on the following:

1.   **Novelty:** The product or service is desired and/or needed and is new and no other competitor is offering a similar value. This could be through developing an entirely new product or service, such as the first mobile phone, or by offering a new service on an existing product, such as the first time a camera was added to a phone.

2.   **Performance:** The product or service has a better performance that appeals to the customer such as being faster, lighter, sharper, etc.

3. **Price/Cost Reduction**: This happens when providing a product or service at a lower cost or providing a product or service that reduces the cost of running business or generates savings.

4. **Design**: The product or service has a superior design that customers prefer.

5. **Brand/Status**: The product or service represents a respected brand that customers prefer or confers a certain status. Customers may pay ten times the price of a car if it is of a certain brand.

6. **Risk reduction**: A product or service that can offer the value of reducing the risk faced by the customers. An innovative insurance scheme is an example of this.

The selection of the value proposition depends on the business an entrepreneur is in and the customer segment being served. Any of the following may be a value proposition for a supermarket:

1. Lowest prices

2. Largest variety of organic goods

3. Free delivery for online shopping

The first value proposition, which is mainly based on cost, is rather weak and undifferentiated. The second and third value propositions are better crafted. They address clear customer segments, the health conscious and the busy working respectively. If no other competing supermarket is offering these value propositions, and if the customer segment addressed is large enough, these value propositions represent a good foundation for the business.

11.2.2 Shareholder Value

Business owners value profit and the increase in value of the shares they own. This happens when the business is well-run, gaining market share, and is able to generate value for its stakeholders. Business owners increasingly value making a positive difference in the world besides making a profit. This generally has a positive impact on the environment, employees, and customers.

11.2.3 Employee Value

Employees represent a key stakeholder and they play a major role in achieving the business objectives. As younger, digital-native individuals join the workforce, they are demanding different things from their careers, and employers need to respond to these demands in order to attract and retain talent.

A recent survey of Millennials (born January 1983 onwards) conducted globally by Deloitte indicated that Millennials expect businesses to care; they want to be innovative, and to be given opportunities to practice leadership. They also want to make a difference. For them, success is not only measured financially but also through how society is improving. Clearly, as businesses employ more Millennials, the needs of these employees need to be addressed.

11.2.4 Suppliers and Partners Value

Suppliers and partners value long term relationships with their customers. Large organisations, such as automotive manufacturers, do run supplier development programmes to ensure that suppliers are able to deliver parts and service of agreeable quality. Nowadays, more and more global coffee chains try to give the coffee bean farmers a fair deal. This is another example of understanding what a supplier values and delivering that. Fair deals are desirable not only by suppliers but also some customer segments that are equity sensitive.

11.2.5 Society Value

Increasingly, businesses are expected to be responsible towards the environment and the societies they operate within. Nowadays, Corporate Social Responsibility (CSR) represents a key element of the operation of the multinational corporations with dedicated departments and budgets.

11.3 Business Model (Entrepreneurial Ecosystem)

The entrepreneurial ecosystem is a description of the environment in which the value proposition is created and delivered. This section is largely inspired by the concept of Business Model Canvas outlined in 'Business Model Generation' by Alexander Osterwalder and Yves Pigneur. While the book provides an excellent representation of the business model, I thought it is necessary to include competitors and the prevailing business climate as essential parts of the entrepreneurship ecosystem. Therefore, the entrepreneurial ecosystem described here is an extension of the business model canvas anchoring it into the prevailing business climate and introducing the competitive pressures to continuously keep competition in mind while creating and delivering value. Different components of the entrepreneurial ecosystem are discussed below.

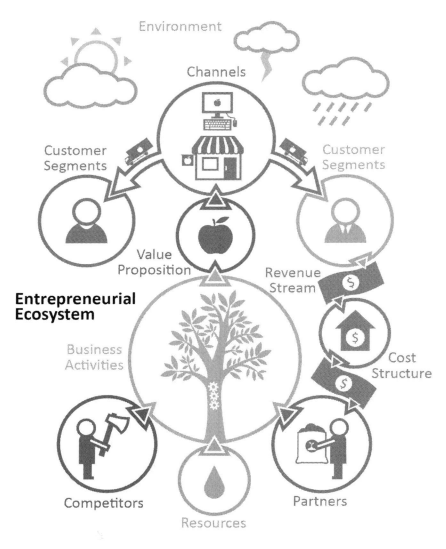

The Entrepreneurial Ecosystem

11.3.1 Customer

Customers represent the centre of the business model and the entrepreneurial ecosystem. They are the reason for the existence of the business and they reserve the final judgement of the value that the business is set up to deliver. In order to address the customer needs more effectively, these needs can be divided based on the type of market they belong to. These include:

1. **Mass market**: The product or service is needed by and available for wide range of customers. This is true for products such as bread.

2. **Niche market**: The value proposition caters to a small segment of the population. An example would be the sale of expensive sport cars or specialised medical equipment.

3. **Segmented market**: An example of this is the airline market. For the same plane that travels from Kuala Lumpur to London, the airlines have economy class customers, business class customers, and first class customers. The market is segmented and a different value is delivered to each segment.

11.3.2 Channels

In order to engage the customers and deliver the created value to them, appropriate channels must be created and kept open. Through these channels a business can create awareness and market its products and services, sell and deliver these products and services, as well as provide after-sales services. Channels will include stores, websites and call centres. Taking a car company for example, the value is delivered to the customer through various marketing activities, making the cars available for a test drive, selling the car, and eventually providing after-sales services and scheduled maintenance. To achieve this, the car manufacturer may use different media as awareness channels - car showrooms for exhibiting the cars and arranging for test drives, and service centres for the maintenance of the cars sold.

11.3.3 Revenue and Cost

Money is the lifeline of any entrepreneurial undertaking. Money can come into the organisation as revenue when goods and services are sold and leave it as cost when payments are made for different expenses. Structuring a sustainable revenue and cost model is a very important part of the business model and the entrepreneurial ecosystem. To turn out a profit, revenue should be higher than the cost:

$$Profit = Revenue - Cost$$

Revenue is generated from asset sales, usage fees, subscription fees, rentals, licensing and brokerage fees. The cost refers to what the business needs to pay in taxes and wages, to acquire different raw materials, products, and services necessary to create and deliver the value proposition.

11.3.4   Resources

In order to realise and deliver the value, a variety of resources are required. These resources represent the ingredients needed to create and deliver the product or service that the business is set to deliver; they include physical, intellectual, human, and financial resources. Entrepreneurs need to secure these resources at a sustainable cost to maintain value creation.

11.3.5   Business Activities

Once the resources necessary to develop the value are secured, specific activities are conducted by the business in order to realise the value. Depending on the nature of the value, these activities can be manufacturing, production, solution provision, and services. If the business is a bakery, then business activities include mixing, baking, and cake decorating.

11.3.6   Partnerships

If you purchase an Apple product online, you will receive your order a few days later through a courier service. Apple does not own the courier service, but rather they have a partnership with it. This way Apple can focus on its core business and allow experts in the delivery business to do their magic. In a similar manner, businesses can form strategic alliances between non-competitors to deliver the value to the customers. These include joint ventures and buyer-supplier relationships. At times, partnerships between competing businesses are formed to control the market supply and demand forces or to lobby policy makers to provide more favourable treatment. Car manufacturers in a country for example, can have partnerships to lobby the government to impose higher taxes on imported cars.

11.3.7 Competitors

Competitors are a fact of life and an integral part of the entrepreneurial ecosystem. Even when an entrepreneur develops a totally new value proposition and creates a new market, it is just a matter of time before competitors appear and challenge the business' position. When analysing the entrepreneurial ecosystem, it is essential to keep competitors in mind. Competitor analysis is essential for entrepreneurial success and the following questions can provide a useful framework for that analysis.

- *Who are your competitors?*

- *What can they offer to your customers?*

- *What competitive advantages do they have over you?*

- *What would you do if you were the competitor?*

11.3.8 Business Environment

Entrepreneurship flourishes by drawing on opportunities available in its environment. The prevailing market, cultural, and legislative conditions can affect the business' success. That is why it is necessary to take these conditions into account when developing a business model. It is essential, for example, to consider whether the economy is growing or shrinking, the price of energy, and the existence of governmental policies to drive the economic activities in a certain direction. When analysing the business environment, the following questions are useful to ask:

- *What is the prevalent market condition?*

- *What legislative system does your business operate in?*

- *What are the opportunities in the environment?*

- *What are the threats to success?*

11.4   Business Plan

A business plan is a useful document that entrepreneurs use to document and communicate the objectives, ecosystems, and aspirations of their businesses to potential investors and financers and other stakeholders such as government agencies.

Business plans are not peculiar to only start-ups. They are written for existing businesses as well, especially when new products are planned or when expansion into new markets is contemplated. A typical business plan has the following sections:

1. **Executive Summary:** The executive summary of a business plan is a one-page summary that highlights the uniqueness of the business, its potential and any interesting facts about it. It is usually written after the completion of the business plan to provide a concise and accurate abstract.

2. **The Company:** This section presents the company (or the team) including the history and track record of the company as well as the organisation chart. The vision, mission, and core values of the company are also featured here.

3. **Business Value:** Here the business value for different stakeholders is outlined. Of special interest and importance is the value proposition for the customers.

4. **Market Analysis:** Critical analysis of the prevailing market conditions and how the value proposition addresses a real need. It also indicates how you plan to market your product or service. It is great if you have sold your products or services and have an existing customer base.

5. **Competitive Advantage:** Outline here any core competencies and competitive advantages that you have. These include any access to information or technology, including patents and trade secrets.

6. **Entrepreneurial Ecosystem and Business Model:** Provide a complete entrepreneurial ecosystem as well as how you plan to make money through defining your business model.

7. **Financial Projection:** Although it is often difficult to provide financial projections, especially for completely new ideas, most of the potential investors need to see some indicative financial projections in business plans.

11.5   CDIO for the Market

Marketing is an essential entrepreneurial activity. It entails knowing the market conditions, the customers' requirements, needs and desires, and communicating the value to the intended customers. There are many books written about marketing, so let us try here to introduce marketing in a slightly different light. Here I wish to present marketing as a holistic process that starts at the Conceiving stage and continues through the Designing, Implementing and Operating stages.

For example, if we want to make a whiteboard pen, we should Conceive it and Design it for the market by keeping the customer needs in mind. So the pen needs to have the right colour, size and weight to give the user a desirable experience. To Implement and Operate it for the market, we have to consider the way we manufacture the pen and how that would impact its marketability. For example we can give the product an edge in the market by using non-toxic inks, utilising only fair-trade raw materials, adopting environmentally friendly ways of manufacturing, manufacturing locally, and providing the workers with good working conditions. There are many customer segments that will not buy the product if it was made in sweatshops or if child labour was used to make it. In summary, marketing with CDIO for the market in mind would require an in-depth knowledge of the customers' preferences, values, opinions and concerns with regards to different issues. Marketing is not an afterthought that takes place through advertising a product or service after it is made.

11.5.1   Market Lifecycle of Products and Services

Products and services, like living things, have lifecycles. Many products or services begin their life's journey by being a novelty. Novel products are often expensive and have limited or poor performance, and only few would be willing to try them or afford them. Think of the first mobile phones, they were expensive, bulky, and had limited performance, at least because the service was not widely available. However, as more and more customers see the potential of the product or service, technology and performance improves and the product or service becomes more common in the market. The lifecycle of products typically ends up by the product becoming a common commodity, i.e., when the product has become commonplace and there is little, or maybe no difference between the offering of different competitors.

In the best-selling book 'Crossing the Chasm', Geoffrey Moore detailed five stages in the lifecycle of a product in relation to its market penetration. When a novel product or service is introduced into the market, few people would be willing to try it. Geoffrey Moore calls this pioneering group the "innovators". It is necessary to point out here that the innovators are different for different products and services. While some of us do not care about the latest gadgets, they might wait in line to try the newest restaurant on the block; while others may be willing to pay extra just to be the first to wear the latest in fashion. The pool of innovators who are willing to try a new product is often very small, approximately 4% of the potential market size. The next group of people who might be willing to try a new product is the "early adopters". Early adopters start using a product after being influenced by advertisements or feedback from the innovators or both. The early adopters represent around 16% of the market. Research shows that most new products' market penetration slows down when it reaches 10% as if they reach a "chasm." Products that progress to reach the "early majority," which represents 50% of the market, need to "cross the chasm." Once the early majority adopts the new product, the price of the product will generally decrease, as more and more people are willing to use the product. This paves the road for the late majority and then the laggards to join.

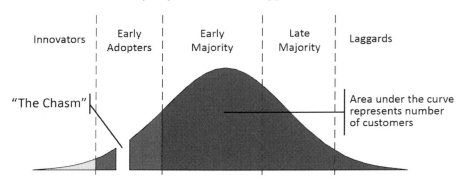

Product Adoption Lifecycle (Source: "Crossing the Chasm" by Geoffrey Moore)

To cross the chasm, a product needs to impress the innovators enough for them to be excited about it and talk about it to their friends and networks. If the innovators become the ambassadors for a product, they can produce enough momentum, together with the early adopters, to cross the chasm.

11.5.2 Market Evolution and Disruptive Innovation

A product starts its life with a certain performance level. As times goes by, often the performance improves. As more providers for similar products come into existence, a market range normally develops with high-end and low-end offerings. With time, the difference between the high-end and the low-end diminishes; that is when the product becomes an undifferentiated commodity.

Taking the smartphone market as an example, there was a time when the iPhone represented the high end of the market with other imitators satisfying the lower end of the market, selling cheaper but with a lower brand and performance offering. Currently the differentiation between various brands is becoming more difficult, and smartphones are almost becoming a commodity.

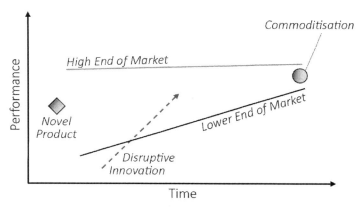

Disruptive Innovation (Source: Innovator's Dilemma, Clayton Christensen)

Clayton Christensen in his book 'The Innovator's Dilemma' presented an interesting theory of what he calls "disruptive innovation." He explains how a new product, service, or company may drive existing prominent market players out of the market and conquer their market share. This happens by addressing market segments that are below the existing lower end of the market. An example of this is the budget airline that started not by competing with the established airlines, but rather by addressing a group of people that the established airlines never saw as customers. They targeted people who travel using trains and buses and offered them a basic low cost, no frills, flying experience. The larger airlines didn't see the new players as a threat at first but as more and more people started travelling with these budget airlines, the larger airlines started losing market share and were compelled to re-evaluate their businesses' models or even go out of business.

11.5.3 Market Penetration

When we examine the market for a certain product (or service) at a given point of its lifecycle, the product (or service) may be anywhere from being a novel product (or service) to being sold in a mature and commoditised market (such as the market for oil, minerals, and other commodities). In between these two extremes, a market often has clear high and low ends of offering with a considerable price differential. To enter the market, entrepreneurs and businesses may select any of the approaches below:

1. Create a novel product.
2. Use disruptive innovation to provide value to a neglected market segment.
3. Differentiate a commodity.
4. Challenge the existing market players by offering better price, performance or customer experience.

Segway: A Product that Did Not Cross the Chasm (Source: Wiki Commons)

Market entry with a novel product is always risky as the product may fail to cross the chasm. An example of a novel product is the Segway scooter. Although cool and highly innovative, the Segway is a commercial failure. More work may be necessary to create an ecosystem in which the Segway will flourish. This can be done by engaging the innovators and the early adopters to fully embrace the technology. Depending on the product being examined, there are various ways to empower the innovators and early adopters to become product advocates and to spread the word around; these approaches include lending the product or giving it away for free, and creating a partnership with these influential groups.

When Nissan wanted to introduce and promote its Electric Vehicle "Leaf" in Malaysia, the company announced a programme where customers can apply to use the car for few weeks. Clearly, this is an attempt to identify the innovators and engage them. Now, the Electric Vehicle is yet to be a common sight on the streets of Malaysia, but the company's move to look for the innovators is definitely in the right direction.

Nissan Leaf (Source: Wiki Commons)

Disrupting existing markets is happening almost on a daily basis. Addressing the needs of what is referred to as the "Bottom of the Pyramid", the billions of people whose needs are not being addressed just because they do not have enough money now; represents the hottest economic opportunity of the 21ˢᵗ Century. Seeing the poor as customers is a very powerful tool for creating products and services that will disrupt existing markets.

Commoditised products can be de-commoditised. The best strategy to market commoditised products is to differentiate your product from the others offered by the competitors. Table salt (chemically known as sodium chloride, NaCl) is a commoditised product. However, a visit to your local supermarket will show many ingenious ways to differentiate this product and charging a premium. While the basic product may cost around RM1 per kg, other variations including extra-fine, extra-coarse, iodine fortified, and sea salt are sold for up to 10 times the basic price. Other manufacturers try to differentiate their product using different ways, these include the use of exclusive and attractive packaging.

Differentiating a Product. Salt as an Example

11.5.4 Customers' Power in a Digital World

In 2010, Adam Brimo was a software engineering and arts (international relations) student at the University of New South Wales in Sydney, Australia when he started to have troubles with his newly purchased Vodafone mobile phone. After the company failed to address his complaints, including dropped calls, poor reception, and no data coverage, all he wanted was to be allowed to exit his contract. To his dismay, Vodafone insisted that he pay a fine before allowing him to terminate his contract. Frustrated by the lack of a working mobile phone, Adam started an online campaign to allow unhappy

Vodafone customers to air their frustrations. I invited Adam to give a lecture as part of my Global Entrepreneurship course and he narrated the whole story to us. His campaign was on a website, which cunningly had the web address www.vodafail.com. "Initially I thought if I have 10 more people who faced the same issues with the service, then maybe Vodafone would listen. However I was amazed to see thousands and thousands of people sharing their complaints and stories online," Adam said. Quickly the campaign was picked up by the local media and the social networks, and soon Vodafone took notice. Adam wanted the campaign to be productive and constructive, so he collated all the complaints in a report that Vodafone could act on. At the same time a law firm saw an opportunity in this and began collecting interest for a class action lawsuit against Vodafone. Vodafone was eventually forced to invest more than AUD 1 billion to upgrade its network and address the complaints. After the lecture Adam delivered to my class, one student tweeted asking whether Vodafone sees Adam as hero or a villain; we were surprised that Vodafone tweeted back that they see him as "an agent of change."

Adam Brimo, Co-founder of OpenLearning

11.6   Lean Entrepreneurship

In the Return on Failure chapter we will discuss the importance of failure, through pushing the limits of what is possible, as a mechanism to seek feedback in order to fine tune the behaviour or the offering, and move on to higher levels of performance. The concept of "Lean Entrepreneurship" or "Lean Startup" represents one of the systematic techniques to yield a high Return on Failure. The concept is outlined in details in the book 'The Lean Startup' by Eric Ries. As entrepreneurs are always in search of repeatable, sustainable and scalable business models, the basic premise of the Lean Entrepreneurship is to quickly (and as cheaply as possible) build products and get them into the hands of customers so that entrepreneurs can measure, learn, and produce even better ideas that will help identify that repeatable, sustainable and scalable business model. The evolutionary Build-Measure-Learn lean entrepreneurship cycle shown below needs to happen in the shortest time possible and with the minimum cost possible.

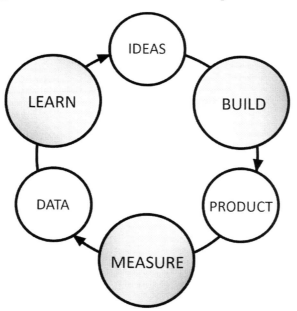

Lean Startup Reduces the Time Taken for the Build-Measure-Learn Cycle
(Source: The Lean Startup, Eric Ries)

Kal Joffres is the co-founder of Tandemic, where he helps NGOs and companies such as Microsoft, Novo Nordisk, and Standard Chartered to leverage the power of open innovation and social innovation to create new products and services. His organisation maintains HATI, the largest database for NGOs in Malaysia. When HATI started to receive requests from people who were looking for volunteering opportunities, Kal realised that there is a need to help match volunteers with NGOs and that was how "do something good" came about. Tandemic spent a considerable amount of time and money building the website *http://dosomething.gd*. "This took a big chunk of my holiday time in France," Kal said jokingly. However after the website was launched, it did not perform as expected! "The website we built was not exactly what the volunteering space needed" Kal told my students when he was sharing his experience with them. "We assumed that few Malaysians are interested in volunteering and that NGOs will be queuing to receive the volunteers. Both assumptions were proven wrong! It turned out that Malaysians are interested in volunteering their time and expertise while the NGOs are apprehensive about how to train these volunteers and how to use their services" he added. Reflecting on this experience, Kal felt that he should have built a much simpler website, spending less time and money, and then check how the volunteers and NGOs would find the website. Although active in the NGOs and volunteering space, Kal admits that without putting a product in the hands of the customer, it is not likely to get the value proposition correct. "Let's face it, your first idea is seldom your best one!" Kal said.

Kal Joffres, Co-founder of Tandemic

Realising that the real need is to develop a solution to help NGOs capitalise on the interest in volunteering as well as to train and manage volunteers, "do something good" used the lean startup principles to roll out a number of successful products including the "Super Volunteers Programme" where "super volunteers" help the NGOs organise and manage the other volunteers.

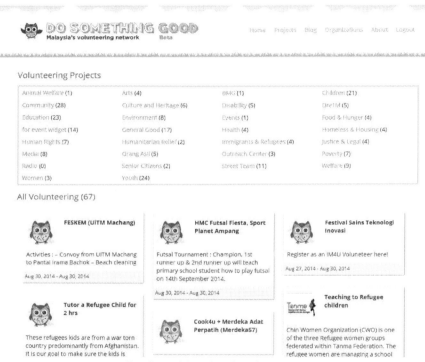

Do Something Good Website

11.7 Funding Entrepreneurship

No chapter about entrepreneurship is complete without a section on how to raise funds to support different entrepreneurial activities. There are a number of ways to get the funds to start up an entrepreneurial activity, including borrowing money from family and friends or convincing some investors to invest in the new business in return for some shares in it.

The latest trend in funding entrepreneurial activities is crowd funding. Enabled by the convergence of the widespread of Internet and the availability of safe payment modes online, crowd funding is making use of the access to millions of individuals around the world to make the dreams of entrepreneurs come true. The premise of crowd funding is very simple, an entrepreneur who has conceived and designed a product and needs money to implement it, can start a crowd-funding campaign online, inviting interested people to pledge money in return for the product when the campaign is successful and the product is manufactured. Popular crowd funding platforms include kickstarter.com, pozible.com and pitchin.my. Besides funding commercial undertakings, crowd-funding is now being used for social entrepreneurial endeavours as well. Community projects are being made possible through the pledges of numerous contributors around the world.

In my opinion, crowd-funding is a very promising way to raise funds for two reasons. First, the crowd-funding exercise is also a form of market study, if the crowd-funding campaign is unsuccessful, entrepreneurs can use this as a very beneficial reality check in order to improve the project for the next iteration. Second, with crowd-funding, entrepreneurs can have a better chance at retaining control over their business ideas without the need to share their business prematurely with those who invest in them.

Thus, crowd funding is the ultimate demonstration of the concept of "Lean Entrepreneurship." It provides both access to funding and the ultimate market endorsement. For a crowd-funding campaign to succeed, customers will need to give their ultimate vote of confidence, namely to pay for a product or service that is yet to completely exist.

In 2014 I offered a free Massive Open Online Course (MOOC) entitled Global Entrepreneurship. A central part of the course, and a requirement for successful completion, is a project where students are required to work in teams and raise funds using crowd-funding. A number of campaigns were successful and the students really had a great time working on their projects. Towards the end of the course, many students told me that the course and the crowd-funding campaign project have strengthened their self-confidence, and that they are now more inclined to attempt entrepreneurial endeavours in the future.

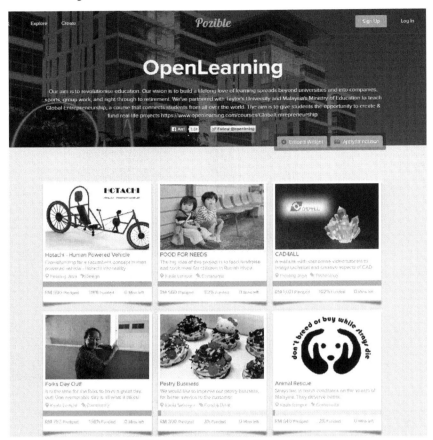

Pozible Crowdfunding Campaigns

Vivian Po is one of my Chemical Engineering students who took the Global Entrepreneurship course. She was inspired by the social entrepreneurs we invited as guest speakers for the course. She ran a highly successful crowd-funding campaign aimed at arranging a day out for the residence of a nearby old folks home. At the end of the course, Vivian posted the following post on the course website. "Being a girl who studied longer than the usual, my life was filled with disappointment and procrastination. I was weak, unorganised and full of negativity. But I've always had a big dream to have my own unique business. I chose this elective course to experience a subject that is out of the engineering core while trying to fix my current financial crisis. To my surprise, this course taught me more than just making money. It taught me how to become a better person and that I'm able to make a difference in people's life while adding value.

Vivian Po, Chemical Engineering Student

"I am truly grateful for how this course was designed as it helped make me who I became today. Although this may sound like a new religion, I personally think it's a two-way traffic. You can make a difference in yourself if you want to and you have to believe in what you are doing. Trying new things won't harm you and who knows, you may be surprised by the outcome. One significant activity that impacted me is the Brain Rewiring. I believe it truly made me a different individual. Not using the 'P' word really made things better, because when you rewire yourself towards positivity, whatever Challenge you are facing transforms into an Opportunity. That Opportunity would bring a new idea, that new idea might be your new direction for a new business, and that business brings you money that you want. See that connection?

"Through a series of inspiring sessions by awesome speakers taught me that business is not only about money. Money would be the second or third prize for a successful entrepreneur. The impact they made, the amount of failures they went through, realising their dreams and how their passion drives them to where they are now is priceless! Even money can't buy the satisfaction and pride they achieved with their own hands. All these indeed changed my perception of business! Aside from the speakers, I really had a great time with my online course mates! Working on a project with team members in different time zone was indeed challenging but it was truly a good experience. From that, I got myself few precious friends!"

196

Entrepreneurship = Innovation = Value Adding

Chapter 11
Entrepreneurship and Innovation

Chapter 12
Return on Failure

*"Try Again. **Fail** again. **Fail** better."*

Samuel Beckett

*"I have not **failed**. I've just found 10,000 ways that won't work."*

Thomas A. Edison

*"I can accept **failure**, everyone fails at something. But I can't accept not trying."*

Michael Jordan

"If you want to increase your success rate, double the failure rate."

Thomas J. Watson

High performance always exists at the fringes of failure! High performers are called so because they are able to achieve things that the others deem difficult or even impossible to accomplish. Think of the athlete who breaks an Olympic record or the scientist who finds a cure for a serious disease. Analysing the success of these master performers in different fields reveal that prior to achieving mastery, producing masterpieces, and high level of performance, these masters have failed again and again, learning from their every failure and using failures as stepping-stones towards success. This is true if we study Albert Einstein, Henry Ford or Michael Jordan.

Many people use WD40 spray to loosen rusted metal parts, but very few know why the product is named WD40. The reason behind the name is that the manufacturer tried 39 failed formulas before reaching the successful 40th one. The Wrights brothers tried more than 200 wing designs and crashed their plane 7 times before being able to fly, and Dyson tried more than 2,000 failing models before he managed to get his revolutionary vacuum cleaner to work, earning him millions in the process.

(From top to bottom) Wright Flyer II, WD40, Dyson Vacuum Cleaner.
(Source: Wiki Commons)

In this chapter we are concerned with the form of failure that could result from pushing the limits and venturing into uncharted territories. It is the failure that may be encountered when novel banking products are introduced, ground-breaking technologies are attempted, and new procedures are experimented with. This failure is a necessary part of learning and should be cultivated and encouraged.

It is an interesting observation that although academic institutions generally have a high level concentration of intellectual capital (high number of highly qualified, smart, and motivated people per unit area), these institutions seem to be unable to sustain the extreme level of innovation and entrepreneurship exhibited by individuals like Steve Jobs and Bill Gates. Both individuals dropped out from very good universities. One of the key reasons for this is that education systems have glorified successes and students are often rewarded to get the "correct" answer. The challenge is that the "correct" answer rarely exists in real life. Often what we have is a range of answers that we need to select from depending on the prevailing conditions and circumstances. Businesses can create innovative performance-driven cultures that appeal to the best employees by avoiding falling in the same trap of glorifying success.

Academic institutions, too, need to encourage students to develop a high performance mindset. A key feature of the high performance mindset is to seek opportunities to learn new skills and capabilities that are beyond what is currently known, and thus pushing and stretching one's limits. This will inevitably lead into some failures, and we need to recognise these as opportunities for growth and feedback on our performance.

12.1   Learning and Failure

As learning new things involves pushing one's limits and venturing into new growth territories, failure is inevitable and is an important step towards success. If we look at the major learning that took place in the life of each of us, learning to walk and learning to speak stand out as major endeavours and successes that changed our lives in profound ways. Now try to imagine, how did you learn to walk? You did not stand up one day at the age of 3 and started climbing up the stairs! The journey would always begin by crawling, then taking clumsy steps and falling, until you gradually mastered the art of walking. The same thing can be said of how you learned how to read, play basketball, swim or drive a car.

Like learning new things, inventing new solutions and providing novel value require venturing into new territories and acquiring new knowledge and skills. The 10,000 times Edison failed prior to succeeding with the light bulb were necessary investments to achieve success. The crucial thing here is to know how to fail well. The key to failing well is analysing each failure and understanding its root cause. Once the root cause of failure is discovered, strategies to improve performance should be developed and executed. This is followed by trying again, testing, and even failing again to get more performance feedback. Small failures during learning have a similar effect to that limited infection has on the immune system; it makes it stronger. On the contrary, preventing the learners from experiencing failure will make them fragile and unprepared, just like an immune system that has never fought germs.

The Performance Growth Model shown schematically in the figure on the following page provides a mental framework to appreciate the role of failure in accomplishing high levels of performance. At the centre of the figure is the expected level of performance, this represents what we can do at a given time and is applicable to individuals or teams. An example of this would be how fast we can perform a certain task, how good we can play a certain sport or speak a given language, or how satisfied our customers are.

If individuals or teams are interested in moving to the next level of performance, beyond what is currently achievable by them, they will need to learn new skills, assume new attitudes and try new things. Clearly attempting to do things differently may result in failure and hence there is an attempt-fail duality denoted by the symbol ($\rightleftharpoons$) at the border between the current level of performance and the next level. The Performance Growth Model can help us see this attempt-fail cycle as a necessary process for building momentum leading towards having enough escape velocity to attain success and moving to a higher level of performance. This understanding is essential to sustaining the motivation as failure may happen again and again, especially when attempting highly challenging objectives. Examining how innovators view this failure is very instructional. It is not a coincidence that Edison said, "I have not failed. I've just found 10,000 ways that won't work." Clearly he managed to convince himself that attempting anew after every failure is the only way to achieve enough momentum to accomplish success and create a new reality, a world that is lit by electricity.

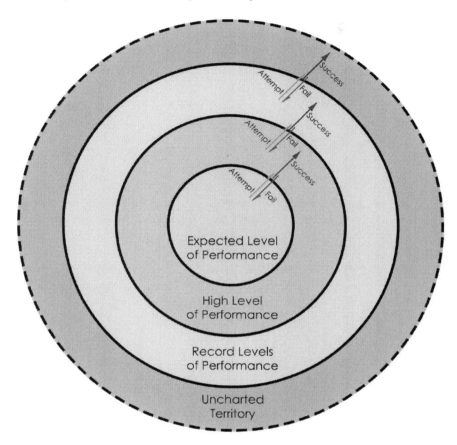

Performance Growth Model

The Wright Brothers successful flight took place in 1903. This accomplishment literally propelled humanity into a new territory, travel through air. Sailing into uncharted territory meant that the Wright Brothers needed to discover the science behind human-powered flight, build and test the technology to achieve that. They operated not only at their own highest knowledge and capability level, but also that of the entire humanity at that time. Crashing their plane 7 times and changing their wing design 200 times, the levels of failures that they encountered were so disheartening that Wilbur Wright declared in 1901: "Not within a thousand years will man ever fly." However, we all know that they prevailed and eventually created history.

It is important to mention here that when we are operating at the border between record levels of performance and uncharted territories (the outer two circles of the Performance Growth Model), building momentum through the attempt-fail cycle may take more than one lifetime or a career before achieving success. Sometimes those who start the cycle may not be around when success is reached. For example, scientists have been working very hard for the past few decades trying to find a cure or a vaccine for AIDS. Thus far this goal has eluded us and we are still in the attempt-fail cycle stage. What we are sure of is that with every failure we learn more about the disease, and we come a step closer to achieving our goal.

12.2   Failing Smart

So now we understand the importance of failure as a step towards success, the question is how do we fail smart? Failing smart will have the following features:

1. Failing smart involves trying new techniques and strategies. These techniques and strategies can be new to us, as we did not practice them at our current level of performance, but are used by other performers who are beyond us. For example, in golf, there is a certain technique needed to swing the club in order to achieve the optimum results. The technique involves an orchestrated series of movements performed by the feet, legs, hips and arms. If you are just starting to play golf, that standard technique will be new to you (although other professional players may be using that technique naturally). While trying to use the technique you will fail again and again until you master it and move to the new level of performance. If you are on the top of your game, any performance enhancement will include some sort of record breaking. Once more, this will require trying new techniques (often never tried before) and failing while attempting to perfect them prior to breaking the record and creating a new level of performance.

2. To be useful, failure should also be as fast and as cheap as possible. For example, if a bank is experimenting with a new technology to replace ATM machines, it is very useful to quickly install this technology in one branch, observe the failures of the system and quickly and correct them with as little cost as possible. We do not

want to roll out the new technology nationwide at a huge cost just to realise that correcting it is going to affect the viability of the bank.

12.3   Failure, Risk and Uncertainty

Donald Rumsfeld, US secretary of defence during the US invasion of Iraq, said that there are two kinds of unknowns, known unknowns and unknown unknowns. Known unknowns are events for which we know their probability of happening; for example, we know how many people die of heart-related diseases each year in a certain population group. When the known unknowns are negative in consequences, they are called risk. There are data available on the probability of the success of new business ventures, such as restaurants. So anyone who wishes to open a new restaurant should know that only 10% of new restaurants survive and make money. When trying ventures with a known risk margin, every effort needs to be made to ensure success, and any failure encountered should help to either reduce the overall risk (through improving the process) of the next attempt to have more likelihood to be within the success zone.

Unknown unknowns refer to events that may happen but we have no idea about the probability of them happening. I am writing these lines while sitting next to the swimming pool where my kids are swimming; there is no way that I can completely rule out the possibility that a meteor hits the exact point where I am sitting. However, the likelihood of this happening is very small (or so I hope). Because such an event, like being hit with a meteor is not a common encounter, we actually do not know the probability of it happening. Such highly unlikely events are called "black swans," a term popularised by Nassim Taleb who wrote a book carrying the same title. Black swans are a manifestation of the uncertainty in the system. The term black swan is derived from the fact that the statement "All swans are white" was thought to be validated by countless observations. It took only one sighting of a swan that is black in Australia to render this statement false. Black swans can be negative, like if a meteor hits where I am sitting now; or positive, like a gold mine is discovered underneath my house.

In order for us to benefit from a positive black swan, it is necessary that we build a system and a culture that encourages tinkering, experimentation, risk taking and yes, failure. This culture should prevail in both the education and business world. The history of black swans is filled with unintended discoveries such as the microwave oven, vaccination, and Viagra. These discoveries were made possible because highly motivated people kept on pushing the limits of our knowledge while keeping an open mind to capture any unexpected opportunity or a positive black swan.

12.4   Education and Encouraging Failure

The importance of failure can never be underestimated in learning, progress and development. Unfortunately, the education system is geared only to celebrate success. Failures are ostracised and students are put in an environment in which risk-taking is not rewarded and often punished. Believing that unless we tinker and take risks, we will never be able to teach others how to do so, we experimented with making failure a requirement for success by requiring each student to report a failure that they learned from while working on their projects. The quality of the failure and the depth of the learning and growth resulting from the failure are assessed for 10% of the final mark of one of the courses. The form that we requested the students to complete is shown on the next page.

Return on Failure

Making mistakes and failing is an integral part of learning. Failures and mistakes can be the result of accidents, miscommunication, ignoring instructions or regulations, or ignoring basic laws of nature. Failures can also be a result of trial and error when the correct answer or the right solution does not exist or has not been discovered yet.

Failures are often a source of very valuable learning. In order for us to reap the full benefit of the failures we encounter and the mistakes we make, it is necessary for us to see failure as an investment that we can seek return over. This form is named Return on Failure and is designed to help you analyse your failure and grow. Let us start!

Complete the sections below. You may expand the space and use diagrams and pictures as necessary

Describe the failure or mistake that you are analysing
(Describe whether the mistake or failure is physical, technical, relational or otherwise. If the failure was done in the course of a trial and error process, describe the cutting edge of technology/knowledge that you are exploring as well. The failure may happen while you are testing a new process or device or while you are trying a new skill. Use pictures, sketches and diagrams if necessary)
Examine what was the Root Cause of the failure
(Ask 5 Why questions starting with "Why did this failure happen?" if the answer is the failure occurred because of "X", then ask "Why did X happen?" and repeat this 5 times. This will yield the Root Cause of the failure)

Are there any other ways that you could have failed to achieve your objectives?

(Here try to predict other ways that failure could have also happened)

Describe how you will use the insight above so that you eliminate or minimise the possibility of failure in the future.

What are the other key learnings from this failure?

Failure is an important stepping stone on the way to success. In order for us to grow, we need to operate at the upper limits of our current performance level and be ready to push these limits and welcome failure as useful feedback for improvement.

Practical Takeaway
Next time you encounter failure, complete the Return-on-Failure form in this chapter so that you can draw the necessary lessons from your failure.

Highlights

Chapter 12
Return on Failure

Chapter 13
Structured P Solving

We have made a pledge not to use the P-word, so the purpose of this chapter is to introduce systematic and structured ways to resolve challenges and realise opportunities. In this context, a challenge can be described as a situation in which the creation and delivery of value is obstructed. A challenge can be viewed as an opportunity waiting to be realised through clearing the obstructed value channels or even discovering new channels that will yield higher value or perhaps entirely new values. This is true if the challenge we are facing is related to health, family, sales, studies, research, manufacturing, or customer services. Imagine that value is being created and delivered through a network of channels, and when some of these channels are obstructed or blocked, a challenge manifests itself. In order to allow the value to flow freely, it is necessary to understand how the channels are obstructed, what obstructed them, and how to clear them or how to create alternative channels. The obstacles that obstruct the value channels can either be real or artificial. Artificial obstacles are often represented by mental models that perceive reality in an incorrect way. These include wrong beliefs and misconceptions. This chapter describes some structured techniques that can be used to solve challenges and clear the value channels.

13.1 Challenge Identification and Formulation

A very important step for resolving any challenge is to accurately and correctly identify and describe it. This may sound intuitive, but you will find it surprising how often people facing a challenge describe it wrongly, inconsistently or even fail to have a description at all. A "Challenge Statement" is a statement that outlines the challenge being encountered. It is written in a very clear and specific language and should identify the *nature of the challenge,* its *impact*, and *how long* it has been affecting the value creation and delivery. Examples of well-defined challenge statements are given below:

"A company is rapidly losing market share to new players in the market. The market share dropped from 25% to 10%. This challenge persisted for the past three quarters and if this continues at the same rate, the company will go out of business in 1 year."

"The machine breaks down frequently causing a down time of 20 hours a week. This results in a delay of delivering customers' orders and the customers' dissatisfaction is on the rise. This has been the case for two months now and if not addressed, it can result in lost business of RM100,000."

"Shoplifting results, on average, in a loss of RM5,000 per week. This represents a ten-fold increase, and it took place over the last six months. If left untreated, this can result in annual loss of RM250,000."

13.2   Root Cause Analysis

After clearly stating the challenge, its impact, as well as its duration, the next logical step is to find the root cause for it. In other words, we need to identify the value channels that are obstructed and what is obstructing them, causing the value creation and delivery to be disrupted; in order to fix the root causes rather than just tackling the symptoms. Just like diseases that infect us, the root causes and the symptoms are often different. Fever, headache and allergic reactions are frequent symptoms to a variety of diseases and an effective cure needs a correct diagnosis in order to target the disease, not only its symptoms. There are a numbers of techniques to find the root cause of a challenge, these include the 5 Whys, 4Ms and the Fishbone diagram. The following sections give a brief description of these techniques.

13.2.1   The 5 Whys

A simple and effective way to find the root cause of a certain challenge is to ask enough whys (normally 5 whys are sufficient) until the root cause is revealed. For example, the challenge faced may be "the machine breaks down frequently," we can ask the "why" questions and a typical sequence may be:

Q1   : Why does the machine break down frequently?

A   : Because the fuse keeps on blowing up.

Q2 : Why does the fuse blowing up?

A : Because the wrong rating was used.

Q3 : Why was the wrong rating used?

A : Because the maintenance personnel are not properly trained.

Q4 : Why are the maintenance personnel not properly trained?

A : Because the training department does not have the right staff to do the training.

For the above challenge, it is clear that the root cause is related to the training and capability of the staff. While it is necessary to fix the broken part of the machine, clearly enhancing staff training and capabilities will ensure that the issue will be resolved and it also ensures that similar issues will not crop up in the future. This may be achieved by developing policies for hiring the right staff, engaging external trainers to train the existing staff, or even using an external vendor to perform the maintenance.

13.2.2 The 4M Method

Obstruction of value creation and delivery can happen because of one or more of the following factors; Man, Method, Machine, and Material. One way of understanding what obstructed the value channels is to study the 4Ms and see how they contribute towards creating and delivering value. These factors are described below:

Man

This refers to the people involved in the creation and delivery of the value. These people need to have the skills, attitudes and capabilities necessary to do their jobs efficiently and effectively. People involved in creating and delivering the value need to be qualified, trained, well compensated and motivated to do their best and ensure that the intended value is consistently created and delivered.

Method

Methods and processes represent the recipe used to create and deliver the value. They refer to the way a patient is examined at a hospital, a haircut is executed at a saloon, a phone call is handled at a call centre, a car is assembled at a factory, a washing machine is repaired at a workshop, and an exam paper corrected at a school. These methods should be well documented, clear, efficient, effective and consistent. A good method should have no bottleneck and should also be free of non-value-adding steps. These methods can be documented using flowcharts and SOPs (Standard Operation Procedures).

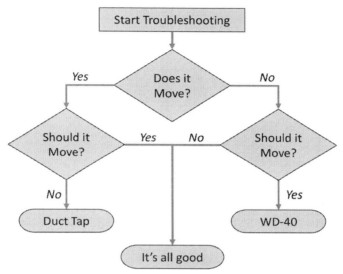

Flow Chart for a Process

Machine

The tools of the trade are a very important part of creating and delivering the promised value. Using the right equipment is key to a successful, sustainable, and economical operation. Using the wrong, unmaintained, and uncalibrated tools and machines can result in a variety of challenges that will obstruct the value creation and delivery.

Material

Materials refer to the ingredients necessary to create and deliver the promised value. If the value is a physical product, then using the right materials to make the product is an important prerequisite to having a successful product and a good customer experience. This concept can be extended to other resources that can be part of making of the product or service even if it is a virtual one.

So let us examine one of the challenges described earlier using this method. The challenge statement is "Shoplifting results, on average, in a loss of RM5,000 per week. This represents a ten-fold increase and it took place over the last six months. If left untreated, this can result in annual loss of RM250,000."

The Use of 4M Method

Element	Questions
Man	Are the personnel well trained?
	Are there enough personnel to man the outlets?
Method	Is there a proper process to control the stock movement?
	Is there a process to prosecute shoplifters?
Machine	Is the technology to curb shoplifting, such as cameras and scanners installed?
Material	Not relevant

13.2.3   Fishbone Diagram

Fishbone diagram represents a visual combination of both the 5 Whys and 4M. The diagram resembles a fish, where the fish head represent the challenge, and different root causes emanate from the spine of the fish. Each major cause may be represented by a bone which can have smaller bones branching from it, representing another layer of causes. To analyse the challenge "A company is rapidly losing market share to new players in the market. The market share dropped from 25% to 10%. This challenge persisted for the past three quarters, and if this continues at the same rate, the company will go out of business in 1 year," the fishbone diagram is given below.

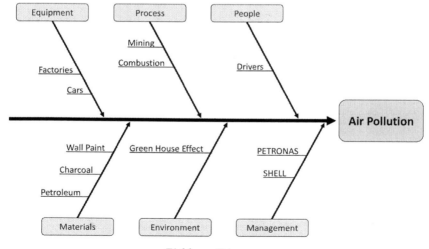

Fishbone Diagram

13.3   Estimation and Quantitative Analysis

One important skill necessary for formulating challenges is the ability to produce meaningful estimations. In order to come up with useful challenge statements, we often need to estimate the impact and duration of a certain challenge. However, coming up with good estimates may not be a straightforward matter. When I ask my students questions like "How many petrol stations are there in Malaysia?" or "How many fish are there in a certain lake?" often the answer I receive is "I have no idea!" This section provides a simple methodology to enable us to develop meaningful estimations of the unknown quantities necessary to solve challenges.

When trying to estimate an unknown quantity, it is necessary to realise that we always have some idea about the range within which the answer exists. For example, for the number of petrol stations in Malaysia, we can always come up with a range that we are confident that the correct answer sits within. Even if you do not know much about Malaysia, there exist a range that you are confident contains the correct number. This could be between 100 and 1,000,000 petrol station. This number can be further refined by trying to get more information about the country, such as the population of Malaysia, the number of cars in the country, or even the length of highways in the country. The key strategy is to first develop the range within which the answer exists. You may take this as an exercise and try to estimate how many petrol stations are there in Malaysia.

Assigning a number or a range to an unknown quantity is a very useful step in converting a challenge into an opportunity. This is a skill that can be developed and honed. Examples of quantities that need to be systematically estimated include: business value, size of the market, satisfaction level of the customers of a competitor and the number of fishes in the sea.

During World War II, the Allies were very interested in knowing the production levels of the German Panzer tanks. The Allies analysts came up with an ingenious way to estimate the number of tanks and the level of production. This was done through studying the serial numbers of the engines and the chassis of the destroyed or captured tanks. Serial numbers are normally sequential and with enough samples, one can work out the sequence in which they are produced. This can be used to infer the numbers of tanks and the levels of their production. To avoid giving out important information, modern military manufacturers are encrypting the serial numbers of the military equipment that may end up in the hands of the enemy.

Learning from this, try to estimate the level of production of 42-inch flat screen TVs of a certain brand (let's say Panasonic). You may use the serial numbers of the TVs on display in a number of supermarkets nearby.

13.4   Final Steps

Now that we explored how to identify and accurately formulate challenges and identify their root causes, the next logical step is to work out a solution that unlocks the value channels and enables it to smoothly flow to the intended recipients. This can be done using the methods discussed in Chapter 4. Methods such as brainstorming, trimming, random entry, and Blue Ocean Strategy can be effectively used to address the root causes of challenges and provide sustainable cost effective solutions for them.

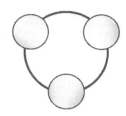

Chapter 14
Engineering a Holistic Education

Holistic / həʊ'lɪstɪk /
Characterised by the belief that the parts of something are intimately
interconnected and explicable only by reference to the whole.

Oxford Dictionary

"Education is not preparation for life; education is life itself."

John Dewey

"The object of education is to prepare the young to educate themselves
throughout their lives."

Robert M. Hutchins

Islam Ali is one of my online students. She is registered for two of my Massive
Open Online Courses (MOOCs), "Achieve Success with Emotional
Intelligence" and "Global Entrepreneurship". Islam is from Jordan and she
graduated with a degree in Mechatronic Engineering in 2007. While I was
putting the final touches on this book, I received a message from her telling
me how much she enjoyed my courses and that she found them useful. I also
got to know that, due to the economic outlook in the region where Islam lives,
and because the traditional industry is not familiar with the field of her study,
Mechatronics, Islam was unable to get a job although she tried for seven years!

"Depression and pessimism were surrounding me. Every small issue
looked like a disaster. I did not think of suicide, but I reached a level that I
was able to understand how people feel when they hate their presence and
commit suicide," Islam Ali said. "Seven years have passed, and each year my

216

birthday felt like a funeral as I just lost another year of my life without achieving anything to be proud of," she added. After joining the MOOCs, Islam started a learning and discovery journey with other students from all over the world. Being a part of a supportive learning community, she immersed herself into the assignments, group works and mental exercises. "When I joined the 'Achieve Success with Emotional Intelligence course,' I was surprised by the immediate and huge amount of happiness and satisfaction I got by watching the first lecture. I watched 7 lectures in the first week! Getting to know my emotions and naming them was really a unique experience to me. Brain Rewiring was so difficult for a pessimistic person like me, but in a week I felt the change. I started to be happy and hopeful almost all the time." Islam said. "Then I decided to join the 'Global Entrepreneurship' MOOC and I was impressed by the fact that engineers are required to master business skills in order to support their work as engineers. Moreover, I realised that being an engineer doesn't mean that I must work in the industry. I can be a teacher with a mindset of an engineer and build generations of successful and happy people. Currently, I'm working on a project with another student from Sudan, whom I met in the MOOC. Our project is a twinning programme that involves students of the same age from Jordan and Sudan." She added.

Stories like the one above repeated a number of times in the free online courses I offered and this got me thinking. How can we provide people with a holistic education that not only prepares them for employment but also prepares them for life, equipping them with the skills and the emotional resilience necessary for living in a fast changing and complex world?

In my mind education, and especially higher education, is not only a requirement to get a job, but more importantly, a character building process through which the students learn how to think and how to learn, enabling them to start their lifelong journeys of self-development, both personal and professional; with the aim of unleashing their potential and the potential of those around them.

I intend to dedicate this chapter to discuss some aspirational thoughts for the future of education and how we can engineer holistic educational experiences. The chapter is also an invitation to discuss and to collaborate.

14.1   Holistic Education

Holistic education can be defined as the provision of three main integrated components, namely Academic Excellence, Emotional Well-being and the development of Lifelong Life Skills.

A Model for Holistic Education

Academic Excellence is probably the easiest among the three to define and measure. However, when it comes to Life Skills and Emotional Well-being, people agree on their importance and differ on how to inculcate and measure them. Some educators even doubt that traits such as emotional intelligence can be taught and learnt. Encountering bright graduates who report that their life was changed after taking some of the MOOCs we offered inspired me to think that approaching the ideal of holistic education might not be that difficult and illusive after all! Here, I am proposing that besides upholding academic excellence, as defined by the norms and standards in the respective academic fields, we can accomplish holistic education by introducing opportunities to grow emotional intelligence (emotional well-being) and entrepreneurial behaviour (lifelong life skills). Although I am driving this point by sharing success stories mainly from the online students, it is useful to say that similar trends were noticed in the on-campus cohort who took these course as well. It is necessary to mention here that entrepreneurship, in the context of our discussion, is defined as the art of adding value and it can be practiced in any aspect of life. As a matter of fact, employers nowadays are increasingly looking for "entrepreneurial employees" who are able to identify needs, and are able create and deliver value.

There is a happy ending for the story of Islam Ali. Realising her passion for education, she is now planning to study a higher diploma in Curriculum and Teaching Methods and attend training at Bright Innovators, a centre that teaches electrical engineering principles for children using simple ways. "I'm planning to have my own school which uses innovative teaching methods and has its own curriculum. This is my vision for the coming 10 years." Islam told me.

She attributes most of the changes of her thinking to her attending the "Achieve Success with Emotional Intelligence" and "Global Entrepreneurship" MOOCs. This supports the hypothesis that emotional intelligence and entrepreneurship can represent the other cornerstones of the holistic education triangle. Beyond the anecdotal story, the initial measurements that we performed on the participants who completed these Massive Open Online Courses (MOOCs) showed very promising results indicating that these traits can be taught and learnt. Links to these courses are available below:

https://www.openlearning.com/courses/Success

https://www.openlearning.com/courses/GlobalEntrepreneurship

Another student who took the "Global Entrepreneurship" MOOC is Caleb Adoh, a final year student at the English Department of the University of Lagos in Nigeria, and this is his story. In an article Caleb wrote online, he described how his university hosted a guest speaker to address the students before they graduate, advising them about life after school. The speaker shared her personal story as an alumnus of the same university and the challenges she faced to prove herself in the job market. "My classmates and I really appreciated her touching story, but she ended her talk sounding pessimistic and negative—she painted a gloomy picture of what it is like to graduate from a Nigerian University, and the supposed fate that awaits us. The hopes and great anticipation of some of my classmates were dashed as she ended her story on a sad note. The very funny part of her talk which I tried to internalise was the fact that she told us she presently runs a fashion store, where she makes clothes, designs and sells them off to potential buyers. She is an Entrepreneur. Unfortunately, she never mentioned anything about entrepreneurship or how we can connect with the entrepreneurial spirit inside of us. Entrepreneurship is not just about being your own boss, as we have been made to believe. Entrepreneurship is starting up something to improve your life and to improve the lives of people around you." Caleb wrote. He went on saying "I have been having meaningful and consistent conversations with my friends at the university and prospective final year students. Only a small percentage of them have thought about starting a venture. Nigerian Universities churn out a great number of graduates at the end of every academic session, but only a very small number of these graduates ever get to the "C" of the CDIO (Conceive Design Implement Operate) of entrepreneurship, talk less of designing, implementing or operating the entrepreneurial idea. There is a type of mental limitation put up by Nigerian graduates and job seekers, and this mental limitation ends up limiting every move they make with regards to having a great career in Nigeria.

"I presently take a Massive Online Open Course (MOOC) on Global Entrepreneurship. There is something we are told to do by the professor daily, it is called Brain Rewiring. You cannot make any meaningful progress as an entrepreneur if your brain is not wired to think in a particular way. It is high time that Nigerian graduates rewire their thinking. We need to think positively and look out for new venture opportunities in Nigeria." Caleb added.

Reading the above, I was thoroughly pleased. Caleb Adoh, an English major graduate, not only captured the essence of the course and its Brain Rewiring component, but also utilised the CDIO framework to analyse the education system and diagnosed what needs to be developed by the graduates in his country. This further supports CDIO as a universal systematic frame of thinking that goes beyond engineering.

14.2   Learning Beyond the School

The American educational philosopher Robert M. Hutchins said "the object of education is to prepare the young to educate themselves throughout their lives." Although he said this half a century ago, this definition of education seems even more pertinent today. In a complex an ever-changing world, the ability to continue learning and the availability of the accessible and relevant learning materials and platforms is paramount. As I mentioned earlier, this chapter is aspirational and I will continue to tell anecdotal stories to drive the point of how important to keep people engaged in lifelong learning journeys.

Our heroine here is Carol Ragsdale, a student registered for the "Achieve Success with Emotional Intelligence" MOOC. Carol lives in the United States and she describes herself as "a 50 year old female, a daughter, a sister, a wife, a mother and a grandmother." She goes on saying "my life is quite simple I surround myself with my family and a few chosen friends but mostly my children. I have four stepchildren, three boys and a daughter, but I am the biological mother of a daughter who is the oldest and a son who is seven years younger." Carol became a single mother at the age of thirty-two after a very difficult sixteen year marriage. She spent the next six years alone raising her children and pursuing a career as a project engineer. Later she met a wonderful man and together they shared their lives and their children, becoming quite a large family. "I was raised by two wonderful parents, my father was a very hard working man and a very present figure in my life, my mother also very hard working with a career devoted to being a wife to my father and a mother to me and my two brothers," Carol remembers. "I was pushed by my parents throughout my life believing that if you work hard, do as you are told and get your education, good things will always fill your life" she said.

"I believed this with all my being and so made this direction the focal point of my life. I would find out on November 11th, 2006 that there was much more to learn about life than what I had been taught during my earlier years. At the age of forty-two I would face my worst night. My son, my baby at the young age of nineteen would lose his life behind the wheel of an automobile. Nothing I had ever been taught prepared me for the devastation I faced. My journey through loss and grief has been a very winding road, one of darkness and aloneness beyond anything I could have imagined." Carol sadly recalls. "The death of my son immediately consumed me. For the next six years I lost my entire identity. The women I was existed no more. Nothing in my life would ever be the same again. Things that brought me joy prior to the death of my son, now brought me only pain and anguish.

"I isolated myself from the world, from my husband, my children and my entire family. I viewed myself as a failure and blamed myself for the death of my son, after all it was my job to keep him safe from harm and I had failed to do this. No one had ever told me that there would be things that would occur in my life that I would be unable to control. I questioned every aspect of my being, including my faith. I could not understand how this could happen to me. My son was such a good boy, his life had just began, he had made it through the tough stubborn adolescent years and was becoming such a wonderful loving, smart, generous young man. I never experienced so much darkness and pain in my life. It was at times as physically painful as it was mentally. I was stuck in a rim between life and death. I would get up after only a few hours of sleep and go to my job and perform my daily task as robotically as can be done. I would have interaction with as few people as possible and even then I would have to fight back my tears.

"Every day I would search the web looking for an answer to how I would ever be able to re-join the world and be part of the living when every aspect of my life felt dead. I took part in grieving classes, both religious and non-religious and became member of organisations to educate parents within our community on the importance of training our children how to drive safely.

"My career started showing signs of deterioration as a result of my growing isolation and lack of communication with my peers who no longer feel comfortable having me as part of their team. My job has been the only thing that I have managed to hold on to. It has become my lifeline and if I can't overcome my grief, I might lose it too. Fear and anxiety flowed through my body as I began wondering how I could possibly overcome yet another loss in my life." Carol said. "One day while I was searching for some magical answer that would pull me from this deep pit of misery and darkness, I ran across 'Taylor's University' on my search bar. The university was offering a free course named 'Achieve Success with Emotional Intelligence.' I was immediately intrigued and without hesitation or my usual research I just enrolled and began taking the course!

"I have to say nothing I had done since the death of my son had reached beyond the surface of my existence the way the first lecture on 'What is Success' did. I felt an immediate rush of excitement and as I began performing the daily Brain Rewiring and My Emotions Today and the other requirements of the course, I couldn't help but feel the change that was now flowing through me. My husband started making comments, as well as my parents and children, as I was, without noticing myself, forcing my way to talk to all of them. I know I still have a long way to go before I can honestly say I have reached the end of my grief. I will always long for my son and wish for him to be here alongside me. But through this one course, I am now seeing the light through which I have also found the laughter and the warmth of love that I thought would never return. I have learned that had I been given certain tools earlier in my life I would have been better prepared and more capable of enduring the struggles associated with my life." Carol added hopefully.

Now Carol plans to reach out to other people to share the importance of holistic education. "Our children should be taught the importance of emotional intelligence as soon as we teach them how to talk! I now believe it is even more important than the physics, science and math that we demand they learn about their emotions and how to manage themselves. At the end of the day the technical knowledge without a true-life understanding is the mechanics without any power" she concluded.

Another story that I wish to share is that of Paul Koba, a young man living in Tanzania. He was one of the first students to join the "Achieve Success with Emotional Intelligence" MOOC. Paul got married in December 2013 and his marriage has a story that is related to the course. "Through this course I watched amazing and life-changing lectures that made me aware of myself, have more self-management and social awareness, and enabled me to develop my relationships with others" Paul said. "I was in love with a young lady, but I could not reveal my love to her despite being together for 3 years" he added shyly. "Finally, I was able to confess my love to her openly. I brought her a ring and I asked the biggest question of my life 'will you marry me?' And she said yes! She told me she loved me too and she can't wait to marry me. I could not believe myself." He added. "I told her that I loved her so much but I was unable to open up and say it. Fear gripped my emotions, I was afraid I was going to lose her. As a matter of fact, I managed to do this after I watched lecture 2 and liked the part that 'All emotions are ok'" Paul added. "She asked me 'where did you get the courage to say that you love me after all these years?' And I replied that I am currently studying at Taylor's School of Engineering a course entitled 'Achieve Success with Emotional Intelligence.' This course is teaching us that all emotions are ok and that encouraged me to share my feelings." he went on saying. Paul reported this on the course website right after the engagement and he generously invited everyone on the course to attend the wedding in Tanzania!

Paul's Engagement

14.3   Mission Zero: A Vision for Higher Education

Thomas Piketty in his book 'Capital in the 21st Century' explored the increasingly unequal global distribution of wealth and the destabilising effects of this trend. The worrying thing is as the demand for higher education increases and the cost of delivering it rises steeply, there is a real danger of aggravating the unequal access to higher education, further worsening the unequal distribution of wealth and jeopardising the opportunity of large socio-economic population groups to escape the grips of poverty.

In order to address this, I am proposing "Mission Zero" which is a vision for higher education that revolves around bringing two variables, eventually, to zero. The first variable is the "tuition fees" that educational institutions charge their students; and the second variable is the "impact on the job market." These two parameters are further explained below. This is done in light of engineering education, but I am sure the model can be extended to other types of educational programmes too.

14.3.1 Zero Tuition Fees

I shall be using numbers that pertain to private education in Malaysia to explain the case here. Malaysia has a thriving private for profit education industry that does not receive financial support from the government. This means the published tuition fees represent a true indicator of the real cost of delivering the educational programmes (plus profit). Currently, educating an engineer in Malaysia cost anywhere between 20,000 to 60,000 USD (over four years). This represents the tuition fees without any living expenses. In order to sustainably reduce the tuition fees and drive them to zero, without resorting to governmental subsidy; the model being proposed here requires rethinking the role of students in their educational ecosystem from being a passive receiver of knowledge to playing a key role in building value-adding systems that can be commercialised and sold for a profit and utilising the revenue to educate the students. The theory is very simple, if we can offer an educational programme for free and still turn out a profit, the model can then be scaled up and deployed anywhere. To achieve this, two convergent strategies need to come together. The first is to view the students as a source of intellectual capital that can be harnessed and exchanged. In a project-based-learning environment, this is straightforward, at least from an intellectual point of view. Our students make hundreds of projects every year, these include racing cars, computer-operated machine and robots. It is not difficult to make the mental leap towards creating projects that the market requires and real customers would be willing to pay for. The second strategy is to forge a partnership with employers with the intention of structuring a "Graduate Tailoring Programme." This entails that the academic institution incorporates the specific requirements of the participating employers who are willing to sponsor a minimum number of students, in the curriculum. This is not a traditional scholarship offered by the employer but rather an investment that comes from staff development budget. The employer will pay it in lieu of the customisation of the educational experience of small batches of students enabling them to be ready to join the sponsoring employer upon graduation. This includes exposing the students to the company culture, business model, software, machinery etc. Done right, this can result in a considerable reduction of training and development cost for the employers.

As educating an engineer in Malaysia costs, on average, 10,000 USD a year (for 4 years), Mission Zero will be achieved when we can earn 833 USD per student per month. This is done through the sale of the students' intellectual and project work, the sponsorship received from the employers, as well as any other source that we can think of as we innovatively address the market needs. It would also be very helpful if academic institutions are able to spin-off successful businesses based on the intellectual work of both students and academics, this way part of the profit can be directed to funding the tuition fees of new students.

14.3.2 Zero Impact on the Job Market

Every year, millions of graduates leave universities and join the ranks of job seekers. While academic institutions are mainly preparing their graduates to compete with those of other academic institutions, I propose the second part of Mission Zero, which is to have an overall zero impact on the job market. The idea is simple; it is about developing entrepreneurial graduates who think of creating jobs, not only competing for them. When a cohort of 100 graduates leaves a university, if 5 of the 100 end up undertaking entrepreneurial activities and employ 95 people, they will offset the 95 jobs taken by their fellow graduates. This way we can ensure a steady supply of jobs for everyone. The 5:95 ratio is just an example.

Clearly, Mission Zero is a journey and as for any journey it starts with a single step. Examples of first steps would be to be able to educate 1 student for free and for the graduates to be able to create 1 job. When it is put in this way, it does not sound as daunting. Achieving Mission Zero requires, above all, the presence of entrepreneurial academic staff, and as this may sound oxymoron, believe me this species does exist! As a matter of fact, I think such academics are yearning for compelling ideas to drive them. I experience that first hand when I interview academics applying for jobs and I see their eyes lit the moment I explained Mission Zero.

14.4   Engineering a Culture Change: Happiness Index

The Kingdom of Bhutan is a small country in the Himalayas. While nations traditionally measure their performance using the Gross Domestic Product (GDP), former king of Bhutan, King Jigme Singye Wangchuck wanted for his country to measure the quality of life in more complete terms using what he called the Gross National Happiness (GNH), striking a balance between the spiritual and material needs.

Last year, on a planning retreat with the senior management team of our university, we were asked to develop strategies to be implemented to improve the experience of our various stakeholders. Inspired by the GNH, I proposed to develop the GIHI (Gross Institutional Happiness Index), a composite index that measures how balanced our experience is. To add a dramatic effect, I also proposed that we rename our title of Vice Chancellor to the CHO (Chief Happiness Officer).

GNH of Bhutan measures 9 domains and they can be adapted to education as shown in the table below. We started collecting the data to create our GIHI and interestingly, few months into the process, the Star newspaper quoted the Second Finance Minister of Malaysia, Datuk Seri Ahmad Husni Mohamad Hanadzlah, saying, "For decades, the progress of the nation has been popularly measured by the Gross Domestic Product (GDP). However, the GDP is inadequate to measure the well-being of the nation and a new approach to measure well-being effectively is required." He said this in his opening remarks before chairing the 2014 Budget Focus Group Meeting on "Developing A Malaysian Happiness Index: Work-Life Balance."

Clearly, academic institutions have a unique opportunity to lead the society in promoting holistic human development, which seems to be a pressing need in this complex and rapidly changing world we are living in.

Domains of the Gross National Happiness Index

	GNHI	GIHI
1	Psychological well-being	Emotional well-being
2	Health	Health and Safety
3	Time use	Time use
4	Education	Learning and Staff Development
5	Cultural diversity and resilience	Cultural diversity and resilience
6	Good Governance	Good Governance
7	Community vitality	Community vitality
8	Ecological diversity and resilience	Sustainability
9	Living standards	Living standards

Outlining the different indicators for each domain is beyond the scope of this chapter and if you are interested in getting more information on our experience measuring the Happiness Index and/or collaborating in this effort, please get in touch.

Epilogue
An Invitation

"If you want to go fast, go alone. If you want to go far, go together."
African proverb

Although the book is entitled "Think Like an Engineer," it is essentially about systematic thinking. It deals with engineering from the wider sense of the word, a word that is a synonymous with ingenuity and the ability to come up with creative and innovative solutions to unlock and enhance value. The book is also concerned with the holistic development of individuals and societies and the fulfilment of the human potential.

It is written as a practical guide to students and practitioners interested in improving how they think and learn. At the same time, the book is also tries to address some of the needs of educators and policy makers by helping them utilise a framework that can sustain a culture change, and the language that can be used to inspire others and communicate the vision.

Brain Rewiring, Opportunity Notes, banning the P-Word, Return-on-Failure, Happiness Index and Mission Zero are some of the initiatives that we use in the ongoing attempt to reimagine education. These initiatives continue to unfold as we speak and my hope is that these experiments will encourage you to implement initiatives at your school, home or workplace and measure their effectiveness so that we can have more alternatives to work with as we go on trying to make the world a better place by investing in the most important capital, the human capital.

I hereby issue an open invitation to everyone to get in touch, comment, share, and discuss any matter in relation to the topics addressed in this book. You can always reach me through my website:

www.thinklikeanengineer.org

Dream Big, Be Different and Have Fun!

Think Like an Engineer

Index

5 Whys, 210, 213

abundance, 62, 101
abundant, 13
achievement, 3, 44, 47, 49, 51, 106,
    120, 125, 168
ACID, 94
adapt-
    adaptability, 44, 49
    adaptable, 23
    adaptive, 83
aeronautical, 5
aerospace, 121
aesthetics, 101
affective, 47, 150-151
affordability, 101
affordance, 94, 100,128
Africa, 61
agriculture, 10, 14
Airbus, 6-7
aircraft related
    aircrafts, 112
    airframe, 7
    airline, 8, 79-80, 180, 186
    airlines, 111, 180, 186-187
    airplane, 7-8
Amazon, 77
analytical, 24
anthro-
    anthropometric, 132
    anthropometry, 132-133
AoN, 167
APA, 148
architecture, 37, 94, 97
artefact, 67, 155
auditory, 19
automobile, 221
automotive, 178

avionics, 7
avoidance, 101
awareness, 2, 13, 30, 45-48, 53-54, 58,
    63, 151, 180

biology, 19-20, 22, 24, 26, 28, 30, 32,
    34, 36, 38, 40, 53, 126, 150
biomedical, 5, 15
biomimicry, 89
Blendtec, 140-141
Blue Ocean Strategy, 78-82, 215
BoM, 98-99, 102
Bozan, Tony 71
brain, 3, 6, 15-16, 19-28, 30-32, 34-36,
    38-40, 42, 44-45, 49-50, 53, 67,
    71, 89, 140, 143, 220
    brainology, 19-20, 22, 24, 26, 28,
    30, 32, 34, 36, 38, 40
    brainstorming, 71-72, 90, 215
    middle, *See* middle brain
    new, *See* new brain
    old, *See* old brain
    rewiring, 30-31, 49-50, 196, 217,
    220, 222, 228
Brimo, Adam 189-190
Business Plan, 182-183

CDIO, 3, 5-9, 20, 30, 34-35, 42, 69-70, 90,
    93, 104, 110, 115-116, 118-120,
    123-125, 145, 159, 175, 184, 220
checklist, 107, 111-114
Christensen, Clayton 186
Cirque du Soleil, 81-82
civilisation, 4, 10, 16, 67, 83
classification, 31-32, 73, 171
classifying, 31-32, 34, 172
cockroach, 31-32

cognitive, 10, 25, 35-36, 43-44, 136, 150-151
    ergonomics, 133
collaborate, 172, 217
    collaborating, 26, 172, 227
    collaboration, 44-45, 58
    collaborative, 150
commodity, 184, 186-187
communication, 12, 15, 26, 44, 55, 58, 63, 118, 137-138, 140-142, 144, 146, 148-152, 154, 156, 158, 168-169, 172, 175, 222
complexity, 5, 36, 110, 160-162
Conceive, 3, 5-6, 8-9, 16, 27, 32, 34, 67-70, 72, 74, 76, 78, 80, 82-84, 86, 88, 90, 92, 115, 118, 137, 155, 175, 184, 220
concrete, 138, 141, 143-144
configuration, 95
    design, 94, 96-97
create, 2, 12, 27-28, 36, 39, 46, 67-69, 78-81, 113, 127, 138, 148, 173-174, 176, 180-181, 187-188, 191, 200, 202, 209, 211-212, 219, 226-227
critical thinking, 3, 9, 36, 44
Crossing the Chasm, 185
crowdfunding, 195
culture, 33, 63, 119, 152, 200, 204, 225-226, 228
cyberspace, 15-16

design
    designer, 6, 53-54, 89, 97-101, 138
    detailed, 94, 98
    optimisation, 99
    overdesign, See overdesign
    process, 54, 94-95, 126-128
desirable, 9, 91, 101, 178, 184
Disruptive Innovation, 186-187
Drucker, Peter 159, 173

Dweck, Carol 20

ecological, 37, 227
economical, 7, 13, 94, 115, 212
    economically, 9, 91, 101, 110
ecosystem, 179, 182, 188, 225
    entrepreneurial See entrepreneurial ecosystem
Edison, 19, 198, 201
efficiency, 14, 89
Einstein, 13, 39, 67, 198
Eliminate, 78, 80-81, 207
Elliot, 25-26
E-mail, 148
emotions, 21, 24, 42, 44-46, 48, 53-54, 142, 144, 149, 151, 217, 222-223
emotional intelligence, 3, 26, 31, 42-46, 48-50, 52, 54-56, 58, 60, 62, 64, 66, 70, 156, 216-219, 221-223
empathy, 42, 53-54, 70, 151
enterprise, 55
entertainment, 16, 81, 142
entrepreneur, 26, 173-177, 181-182, 187, 191, 194-196, 220
entrepreneurial, 173-176, 180, 182, 184, 193-194, 218, 220, 226
    ecosystem, 178-183
    -ism, 44
entrepreneurship, 9, 26, 173-176, 178-180, 182, 184, 186, 188, 190, 192-196, 200, 216-220
    Global, See Global Entrepreneurship)
EPIC Homes, 55-57
ergonomic, 26, 155
    ergonomically, 133
    ergonomics, 126, 128, 130, 132, 134, 136
ERIC, 78-80, 191
evolution, 10, 30, 85-88, 115, 153-154, 186

evolutionary, 3, 9, 94-95, 191

Facebook, 55, 83, 88, 131, 157
failure, 2, 7, 29-30, 37, 59, 61, 100, 115,
    128, 145, 188, 191, 196, 198,
    200-207, 221
    Return on, *See* Return on Failure
feasibility, 160
    feasible, 9, 91, 101, 160
feedback, 28-30, 96-97, 126, 128-130,
    185, 191, 200-201
financers, 182
fishbone diagram, 210, 213
Forming, 21, 40, 104, 154-155
FMRI, 21

Gantt Chart, 162-164
GDP, 226-227
GIHI, 226-227
Gladwell, Malcolm 27, 138
Global Entrepreneurship, 219
GNH, 226-227
GNHI, 227
Goleman, Daniel 42-43, 45-46, 49, 53
Gross Domestic Product, *See* GDP
Gross Institutional Happiness Index,
    See GIHI
Gross National Happiness, *See* GNH

habits, 9, 20, 22-23, 27-28, 32-35, 50,
    173
habitually, 137
happiness, 29, 34, 38, 45, 58, 174, 217,
    226-228
hardware, 20, 98, 104-106
hardwired, 19-20, 24
    hardwiring, 27
HATI, 192
holistic, 3, 12, 36-37, 42, 44, 184, 216-
    217, 227-228
    education, 42, 44, 216-219, 223

Human Centred Design, 126, 128, 130,
    132, 134, 136
humankind, 12-13, 69

ideation, 69, 71, 74, 76, 89-90
IDEO, 126
IKEA, 111, 149-150
Implement, 3, 5, 8-9, 16, 27, 34, 72,
    104, 106, 108, 115, 137, 155,
    161, 175, 184, 194, 220, 228
    implementation, 7, 104, 113, 118,
    163, 175
increase, 51-52, 71, 78-80, 85, 100, 159,
    168, 177, 198, 210, 212
infrastructure, 16, 115
innovation, 11, 14-15, 26, 82, 88, 173-
    174, 176, 178, 180, 182, 184,
    186-188, 190-192, 194, 196, 200
    Disruptive *See* Disruptive
    Innovation
innovative, 14, 26, 49-50, 56, 72, 78, 88,
    111, 174-175, 177-178, 188, 200,
    219, 228
innovator, 173, 185, 186, 188, 201, 219
INSEAD, 78
inspiration, 89
    inspirational, 29, 58-59
integrated, 7, 42, 44, 90, 93, 95, 106,
    151-152, 176, 218
    design, 94, 98
intellectual, 9, 34, 53, 81, 91, 98, 145,
    181, 200, 225
intelligence, 43, 67, 106, 137
intention, 77, 225
    intentional, 4, 22-23, 27, 36, 38, 69,
    127
    intentionally, 6, 50
IQ, 43-44

Joffres, Kal 191-192

Kahneman, Daniel 35, 134-135
Kelly, David 34, 126
kickstarter, 194
Kim, W. Chan 78
Kohlieser, George 63

leader, 13, 29, 34, 55, 58-61, 71, 117,
    120, 146, 152-153, 173
    leadership, 55, 58-59, 121, 154-155,
    178
Lean
    Entrepreneurship, 191, 194
    Startup, 191, 193
learner, 16, 21, 28, 151, 201
lifecycle, 3, 184-185, 187
logbook, 145/

management, 5-6, 12, 57-59, 81, 104,
    110, 124-125, 156, 159, 161-164,
    169-170, 176, 226
    Managing Projects for Success,
    159-160, 162, 164, 166, 168, 170,
    172
Mann, Darrell 71
manoeuvrability, 164
manufacturability, 91, 98, 100
marketability, 184
Massive Online Open Course, 16, 43,
    55, 64, 173, 194, 217, 216-221,
    223.
mastery, 23, 27-29, 69, 138, 174, 198
Mauborgne, Renée 78
measurement, 43, 107, 132-133, 219
media, 13, 34, 61, 128, 165, 180, 190
medication, 37
mentor, 30
Mesopotamia, 10
methodology, 82, 147, 150, 214
Middle Brain, 24, 44, 46, 142
mindset, 2, 5, 19-20, 26-30, 36, 38, 42,
    69, 173-174, 200, 217

mission, 50-52, 58, 65, 183
Mission Zero, 224-226, 228
MOOC, See Massive Open Online
    Course
multidisciplinary, 5, 104, 137, 155
myelin, 22-23, 27, 30-31, 50
    myelinate, 26
    myelinated, 23
    myelination, 23, 31

Network Diagram, 166-167, 170
neuron, 21-23, 27-28, 30-31, 50, 53
neuroscience, 22, 25
New Brain, 24-25, 32, 44, 141
NGOs, 191-193
Norming, 154-155

Oei,John-Son 55-56
Old Brain, 24, 44, 140
OpenLearning, 190, 219
Operate, 3, 5, 8-9, 11, 16, 27, 29, 34,
    100, 102, 110-116, 118, 120, 122,
    124, 127, 130-131, 137, 151, 155,
    161, 175, 178, 182, 184, 220
optimisation, 98
    optimise, 106
optimism, 49-50
    optimistic, 20, 174
optimum, 115, 203
Orang Asli, 55-56
organisation chart, 152-153, 183
outliers, 27
overdesign, 100

Panasonic, 215
Panzer, 214
paradigm, 29, 69-70, 174
passion, 2, 196, 219
PDM, 167

performance, 20, 27, 29-30, 81, 89, 94, 106, 156, 170, 173, 176, 184, 186-187, 191, 198, 200-203, 226

Performing, 15, 21, 23, 27, 42, 47, 50, 70, 78, 87, 97, 113, 133, 154-155, 168, 222

personalisation, 16, 88

Picasso, 138-139

Piketty,Thomas 224

PitchIn, 194

Polaroid, 97-98

positivity, 196

Pozible, 194-195

presentation, 143-144

project based, 122, 225

    Project-Based Learning, 2

proposal, 34, 144, 162, 164

Random Entry, 74-77, 215

recyclability, 98

recyclable, 101

Reduce, 9, 15, 78, 80, 85-86, 90-91, 168, 204, 225

redundancy, 100

relationship, 42, 45-47, 50, 52, 58,122, 131, 142, 152, 161, 178, 181, 223

    management, 45, 58

reliable, 7, 124

    reliability, 90-91, 100, 107

renewing, 47

reptilian, 24

requirements, 7, 17, 69-70, 83, 94-95, 97-99, 106-107, 115, 159, 169, 184, 222, 225

resilience, 217, 227

    resilient, 42, 174

Return on Failure, 191, 198, 200, 202, 204, 206

revenue, 81, 180-181, 225

rewire, 196, 220

    brain, *See* brain rewiring

rewired, 49

rewiring, 27

risk management, 168

Root Cause Analysis, 210

Rumsfeld, Donald 204

SaniShop, 61

sanitation, 14, 61, 175

satisfaction, 159-161, 169-170, 196, 214, 217

scalable, 191

Segway, 187-188

Self

    Management, 45, 48-49, 223

    Assessment, 47

    Awareness, 42, 46

Shakespeare, 19

shareholder, 167-177

Sim, Jack 61-62

simulation, 98, 138, 143

Sinek, Simon 29

Social Awareness, 45, 53, 57, 63, 223

SOPs, 212

Stakeholders Management, 170

stimuli, 19-20, 24, 26, 43-44, 48-49

Storming, 54, 154-155

subsystem, 7, 37, 94-95, 97, 107

subtasks, 166

SUCCES, 138

SUCCESS, 138

    Simple, *See* simple

    Unexpected, *See* unexpected

    Credible, *See* credible

    Concrete, *See* concrete

    Emotional, *See* emotional

    Stories, 143

    Simulation, *See* simulation

success, 2, 20, 28-30, 34, 38, 43, 45, 53, 59, 64, 70, 118, 124, 142, 144, 147, 150-151, 154, 156, 160-161,

169-170, 172-173, 178, 182, 198,
    200-205, 216-219, 221-223
SWOT, 47-48, 58
System Architecture, 94-96
systematic, 3-4, 9, 34-35, 69, 90, 124,
    191, 209, 220, 228
    systematically, 5, 44, 90, 214
systemic, 36

tactile, 19
Taleb, Nassim Nicholas 204
Tandemic, 191-192
Taylor's Racing Team, 107, 116-121,
    151
teamwork, 7, 26, 45, 58, 137-138, 140,
    142, 144, 146, 148, 150, 152,
    154, 156, 158
Tesla, 4
Trend Recognition, 82
trimming, 77, 215
Tuckman Model, 153-154
Twitter, 88

uncertainty, 75, 204

value, 3, 5, 9, 12, 15, 27-28, 49, 52, 69,
    78-81, 90-91, 93, 124, 152, 156,
    168, 170, 173-184, 187, 192, 195,
    201, 209-212, 214-215, 218-219,
    225, 228
verification, 106
viability, 204
viable, 9, 52, 55, 91, 101
Viagra, 205
vision, 29, 50-51, 58-60, 183, 219, 224,
    228
Vodafail, 190
Vodafone, 189-190
Vujicic, Nick 68

Wagner, Tony 44

Warner, Jim 46
Work Breakdown Structure, 162, 164-
    165
WD, 198-199
wellbeing, 218
WMSDs, 133
WTO, 61

38030889R00150

Printed in Poland
by Amazon Fulfillment
Poland Sp. z o.o., Wrocław